BOOK 3

Taotie Folding

Dr. Keh-Ming Lu

Book 3 - Taotie Folding
Copyright © 2022 by Dr. Keh-Ming Lu

Control Number ISBN
Paperback: 978-1-77419-121-7
eBook: 978-1-77419-122-4

All rights reserved. No part of this book may be reproduced or transmitted, downloaded, distributed, reverse engineered, or stored in or introduced into any information storage and retrieval system, in any form or by any means, including photocopying and recording, whether electronic or mechanical, now known or hereinafter invented without permission in writing from the publisher.

Disclaimer: This is a work of nonfiction. No names have been changed, no characters invented, no events fabricated.

To order additional copies of this book, please contact:

MAPLE LEAF PUBLISHING INC.
www.mapleleafpublishinginc.com
3rd Floor 4915 54 St Red Deer,
Alberta T4N 2G7 Canada
General Inquiries & Customer Service:
Phone: 1-(403)-356-0255
Toll Free: 1-(888)-498-9380
Email: info@mapleleafpublishinginc.com

Introduction

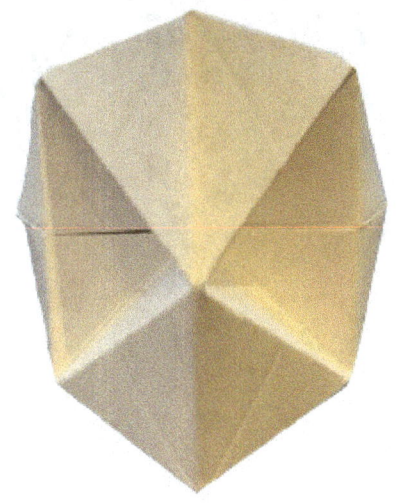

Introduction (1/3)

Taotie (饕餮) model has a head but no body. The model comes from paper folding instructions are available in the form of a crease pattern or CP.

Introduction (2/3)

Among paper material, we are employed a thicker paper creased into polyaboloes P64.

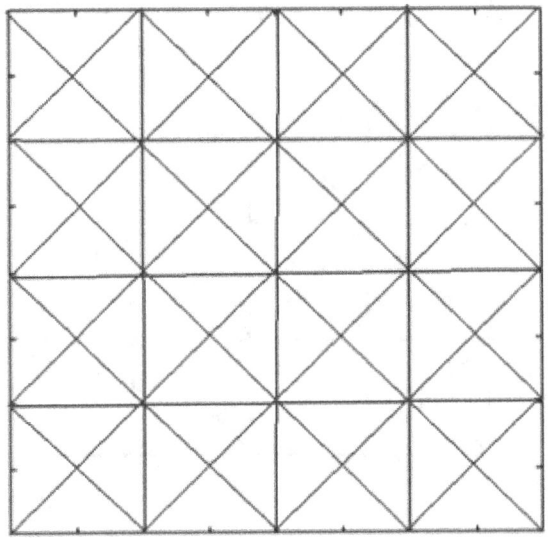

Introduction (3/3)

Throughout the Book: Taotie Folding, we have two types of figures, such as an example: Crease Pattern, CP082 and its Taotie Model, T082.

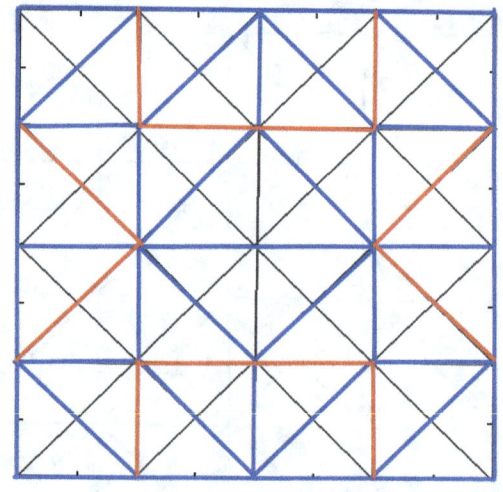

CP082

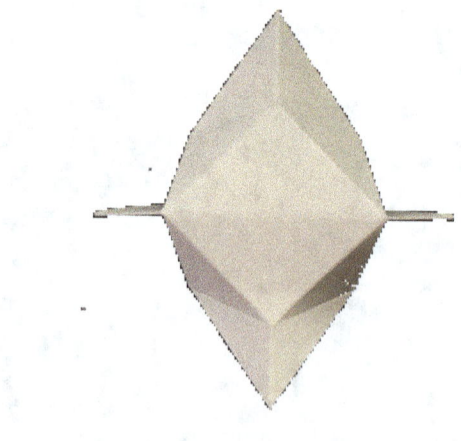

Crease Pattern (1/4)

The crease pattern is defined by four types of creases: Ridge, Ravine, Hinge, and Cave creases. Ridge crease is a fold or a crease that extends inward from a polygon corner or a node in blue ink.

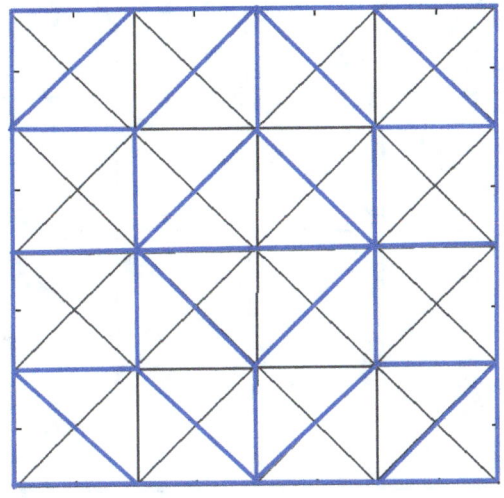

CP082 Ridge

Crease Pattern (2/4)

Ravine creases is a fold or a crease that extends outward from a polygon corner or a node in red ink. Ravine creases are creases that run among Ridge creases from the node.

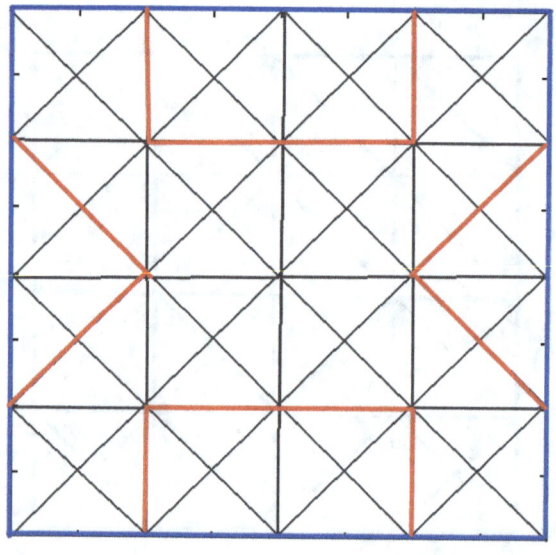

CP082 Ravine

Crease Pattern (3/4)

Hinge creases are a pair of perpendicular Ravine creases coming from a node. A Hinge crease is a line around which a flap can rotate. Hence the name. It looks similar to a door hinge.

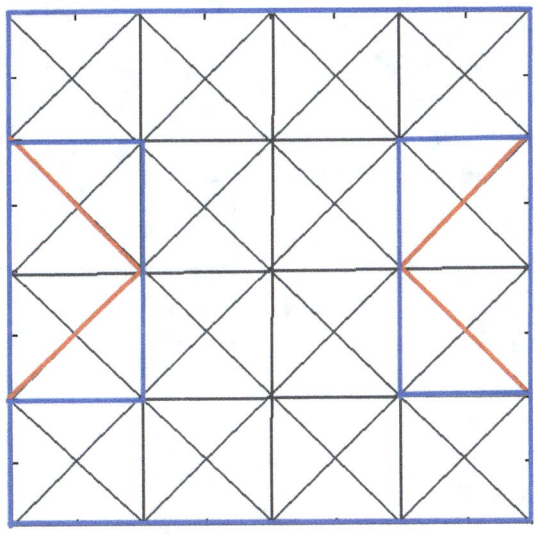

CP082 Hinge

Crease Pattern (4/4)

Cave creases are a cave set of five Ridge creases in blue as a bridge or cover and three Ravine creases in red as a cave floor. As an example, please note that CP092 of T092 has two Cave creases.

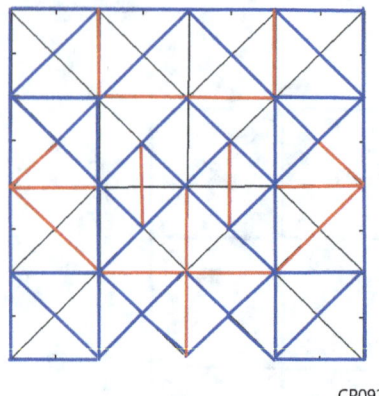
CP092

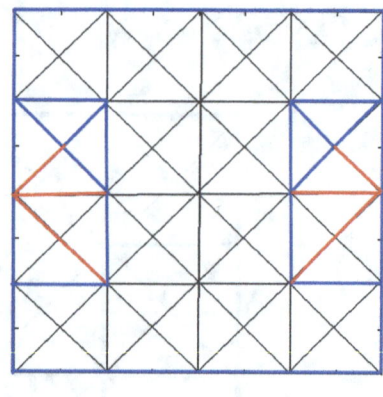
CP092 Cave

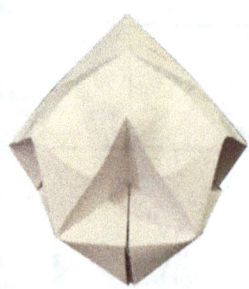

Folding Paper (1/2)

Taotie folding in this book will need scissors to cut off a number of triangles from P64. The folding paper P64-2n if cut off 2n triangles such as P48, P52, P54, P56, P58, P60, P62, P64.

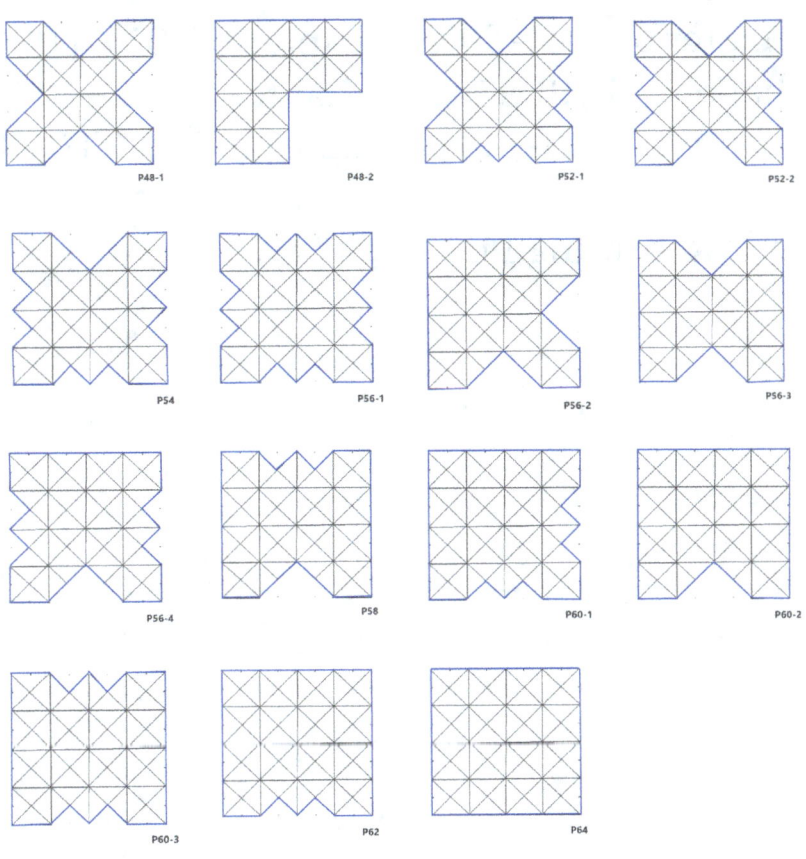

Folding Paper (2/2)

This Book contains 150 Taotie Models starting from T001 through T150. They are divided into groups and described in ten chapters: Chapter 1 T001-T015; Chapter 2 T016-T030; Chapter 3 T031-T045; Chapter 4 T046-T060; Chapter 5 T061-T075; Chapter 6 T076-T090; Chapter 7 T091-T0105; Chapter 8 T106-T120; Chapter 9 T121-T135; Chapter 10 T136-T0150.

Prof. Keh-Ming Lu

T001-T015

T001 (1/3)

The Model T001 is a cube. The Crease Pattern 001 or CP001 of Model T001 has only eight Ridge creases.

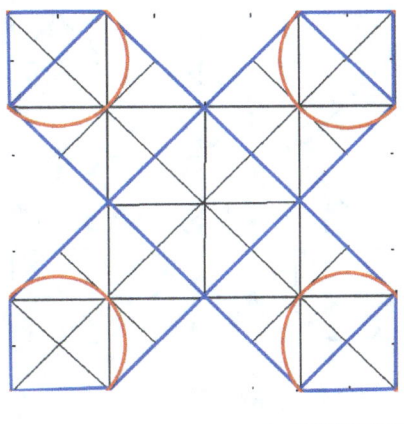

CP001 Corners

T001 (2/3)

There are four Ridge creases at each of four corners.

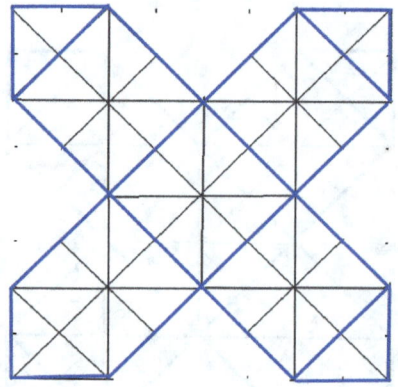

CP001

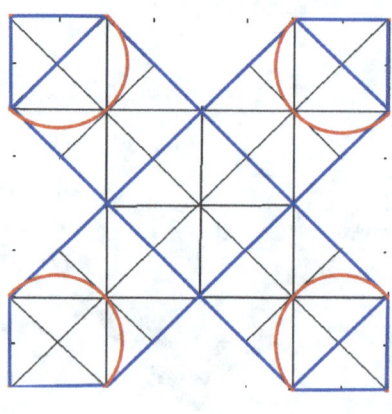

CP001 Corners

T001 (3/3)

Assign numbers 1, 2, 3, and 4 at four corners. The eight triangles of the bottom of cube is now a base that can not be seen.

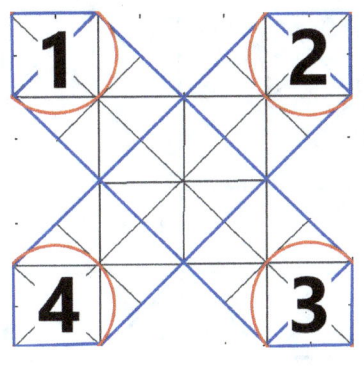

CP001 Corners #

T002

Model T002 has 2 sets of triangle-diamond connected and 4 triangle caves.

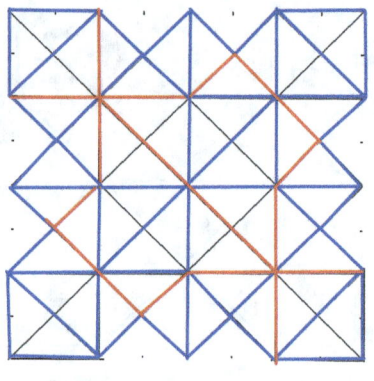

CP002

T003

Model T003 has 2 sets of triangle-diamond connected and 4 diamond caves.

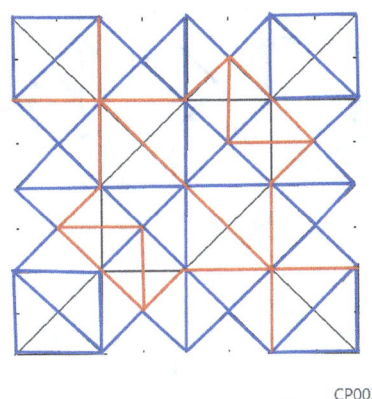

CP003

T003

T004

Model T004 has triangle-diamond connected and two diamonds in one side while 2 triangles and 2 diamond caves in other side.

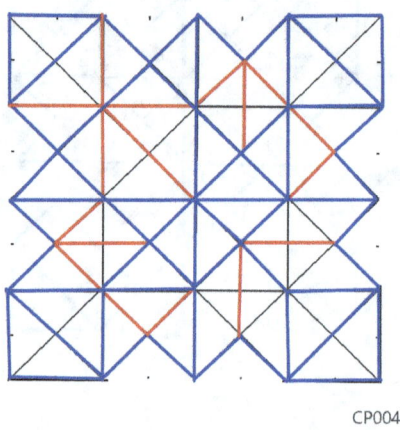

CP004

T004

T005

Model T005 has 1 triangle-diamond connected cave and 2 outward pagodas in one side while 2 triangles and 2 diamonds in the other side.

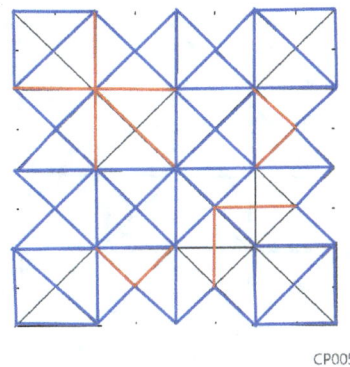

CP005

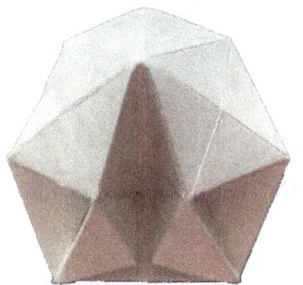

T005

T006

Model T006 has 1 triangle-diamond connected cave and 2 pagodas inward in one side while 2 triangles and 2 diamonds in the other one side.

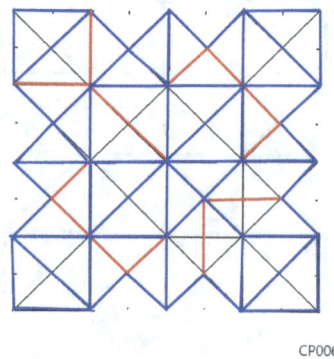

CP006

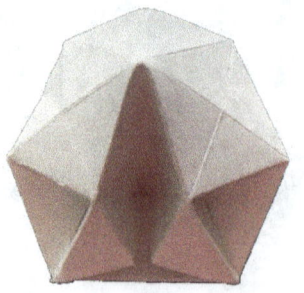

T006

T007

Model T007 has 1 triangle-diamond connected cave and 2 pagodas inward in one side while 2 triangles and 2 diamonds in the other side.

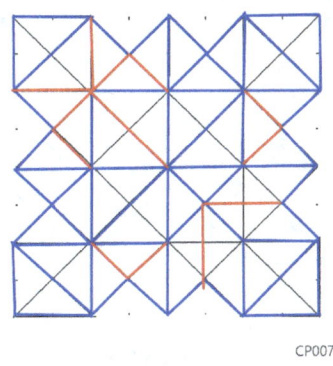

CP007

T007

T008 (1/2)

Model T008 has 4 trapezoid cave compartments.

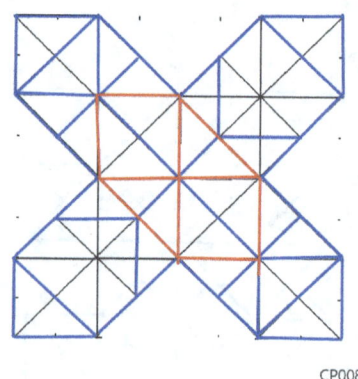

CP008

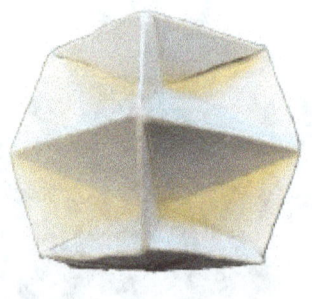

T008

T008 (2/2)

The FP008 has eight Ridge creases and eight Ravine creases (two from two pairs of edges).

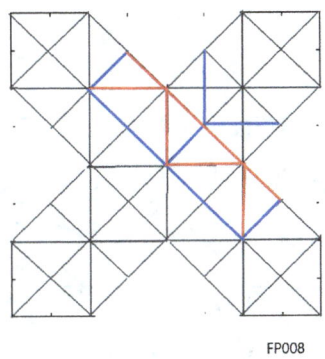

FP008

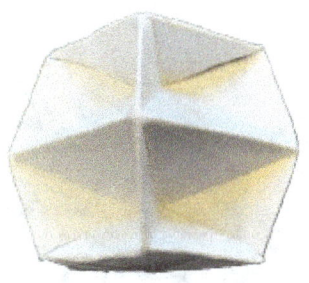

T008

T009

Model T009 has four diamond caves. A pair of diamonds at each side. Top view of T009 looks like the roof of a hanger.

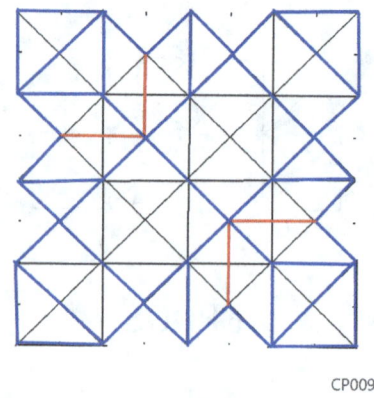

CP009

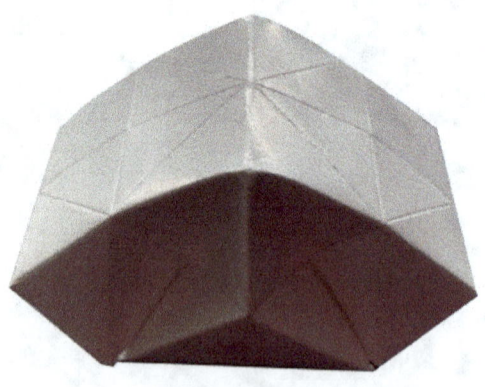

T009

T010

Model T010 has four trapezoid Niches. Niche is a shallow recess in a wall to display a statue. A pair of Niches have two pagodas outward while the other pair of Niches have two pagoda inward.

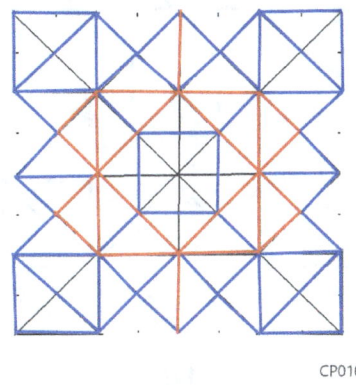

CP010

T010

T011

Model T011 has four Trapezoid Niches. Each has a pagoda outward in the left side while a pagoda inward in the right side.

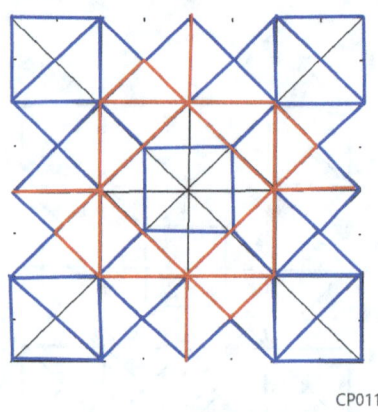

CP011

T011

T012

Model has four Trapezoid Niches. Each has a pagoda outward in the right side while a pagoda inward in the left side.

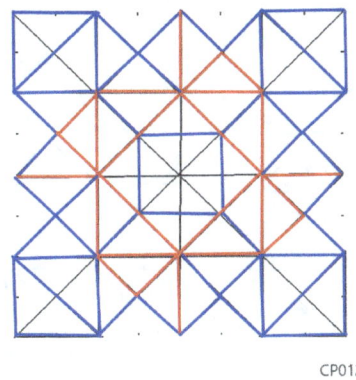

CP012

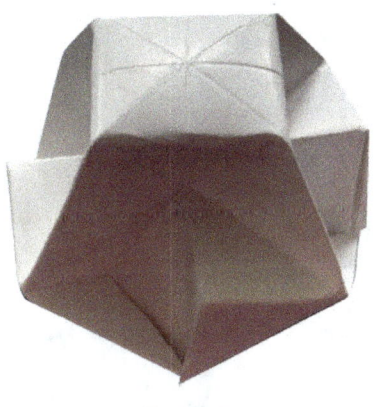

T012

T013 (1/3)

The Model T013 is one of popular Models. The CP013 of Model T013 has 20 Ridge and 4 Ravine creases.

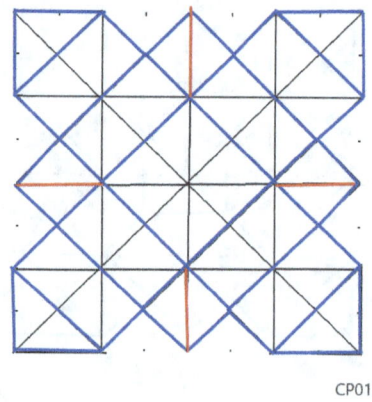

CP013

T013

T013 (2/3)

A Hidden Cube consists of a T001 and 6 x T013. It can be constructed by inserting a combined modules: T001 +T013 into another combined module 5 x T013 There is a cube hidden inside.

Hidden Cube

The Model T013 is an essential module in Four Dice: D6, D10, D14 and D18.

T014

The Model T014 is also one of popular Models. The CP014 of Model T014 has 28 Ridge and 16 Ravine creases.

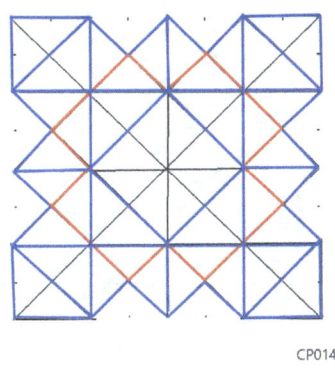

CP014

T014

T015

The Model T015 is also one of popular Models. The CP015 has 26 Ridge and 8 Ravine creases.

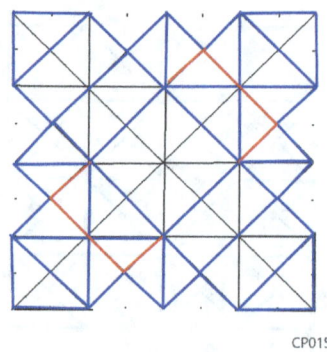

CP015

T015

T016-T030

T016

Model T016 has four compartments. Each one has two pagodas outward. We call it a Niche. Niche is a shallow recess in a wall to display a statue.

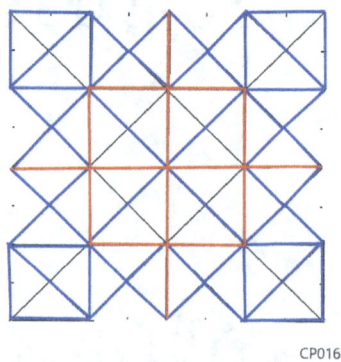

CP016

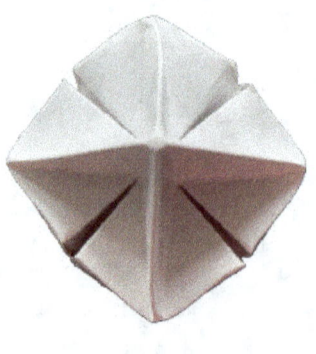

T016

T017

Model T017 has two Niches in the front and a bar of one side of T013 in the back.

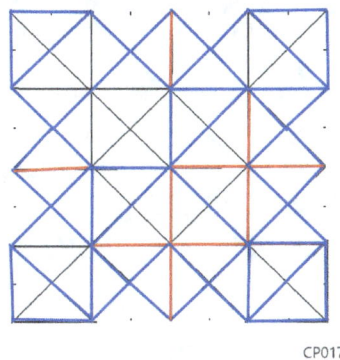

CP017

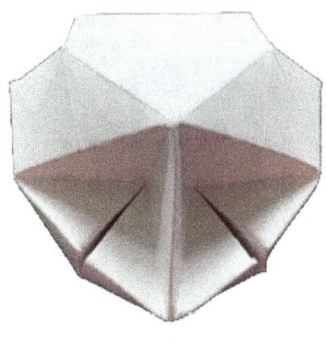

T017

T018

Model T018 has four pentagon compartments each with a pagoda outward.

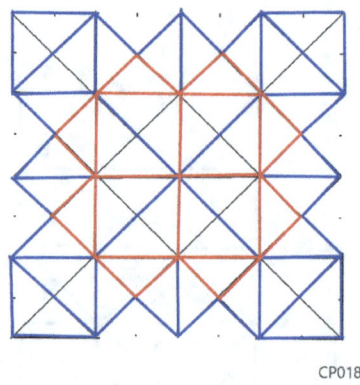

CP018

T018

T019

Model T019 has two Niches in the front each with two pagodas inward and two diamonds and two triangles in the back.

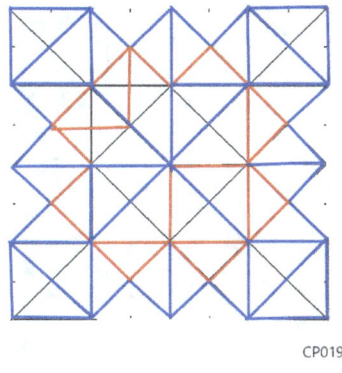

CP019

T019

T020

Model T020 has two Niches in the front each with two pagodas inward and a big Niche with two pagodas inward in the back.

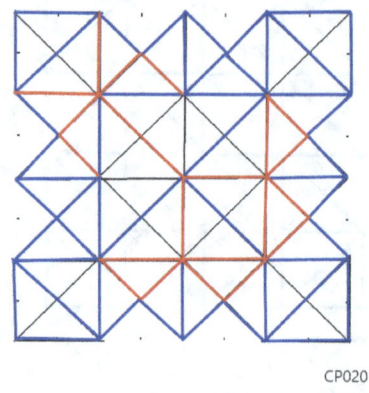

CP020

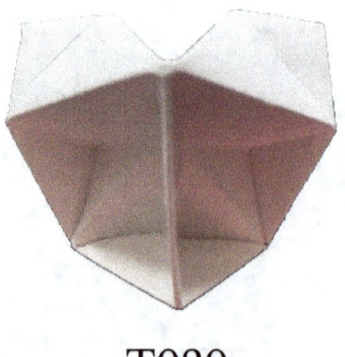

T020

T021

Model T021 has a Niche with two pagodas inward in the front.

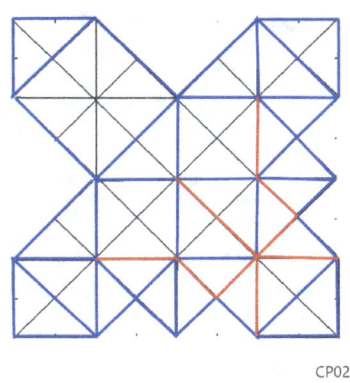

CP021

T021

T022 (1/3)

The Model T022 is one of popular Models. It has four pentagon compartments each has two connected diamonds inward.

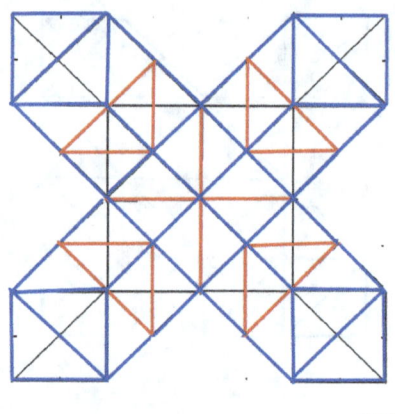

CP022

T022

T022 (2/3)

Sun Flower consists of Flower Head 6 x T013 and Flower Petal 8 x T022.

Sun Flower

T022 (3/3)

Spike Ball consists of Ball 6 x T013 and Spike 8 x T022.

Spike Ball

T023

Model T023 has four pentagon compartments each has two separated diamonds inward.

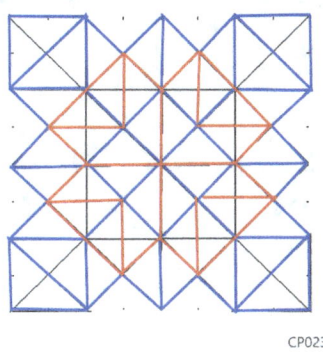

CP023

T023

T024

Model T024 is two diamonds and two triangles in one side and the same in the other side.

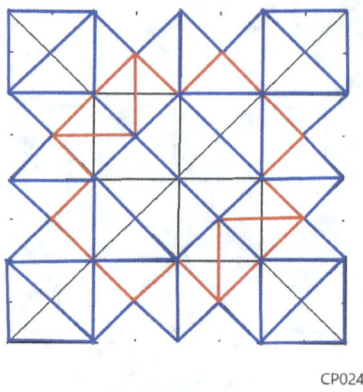

CP024

T024

T025

Model T025 has two Niches in the front and four triangles in the back.

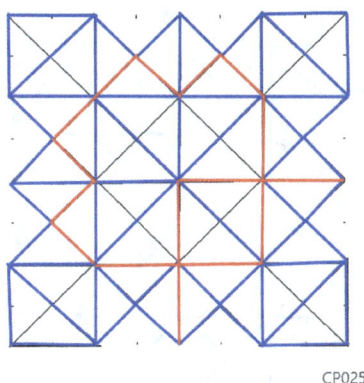

CP025

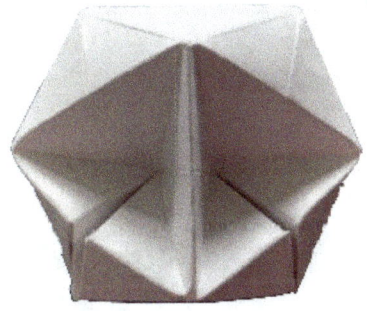

T025

T026

Model T026 has two Niches in the front and two diamonds and two triangles in the back.

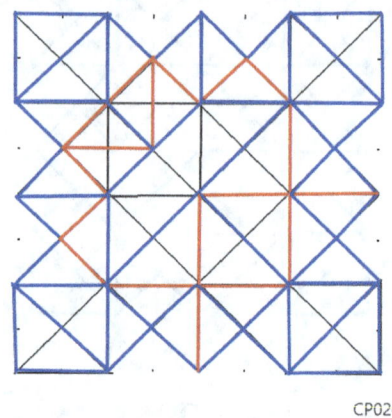

CP026

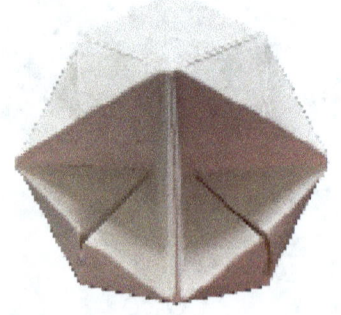

T027

Model T027 has two pentagon compartments with two pagodas in the front and two diamonds and two triangles in the back.

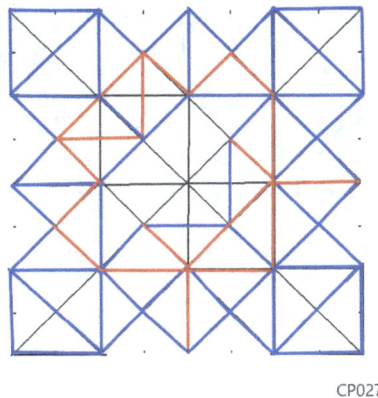

CP027

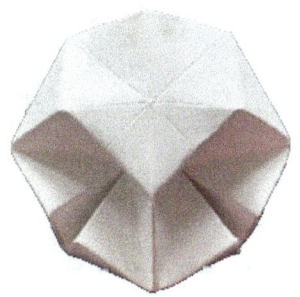

T028

Model T028 two diamonds and two triangles in the front and four triangles in the back.

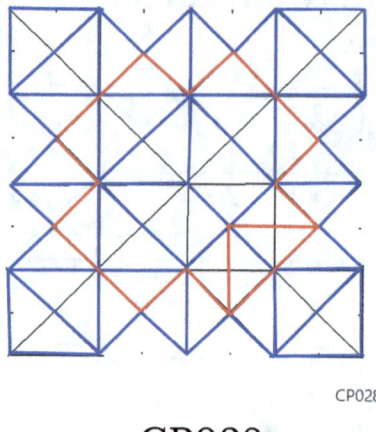

CP028

T028

T029

Model T029 has four pentagon compartment each with one pagoda.

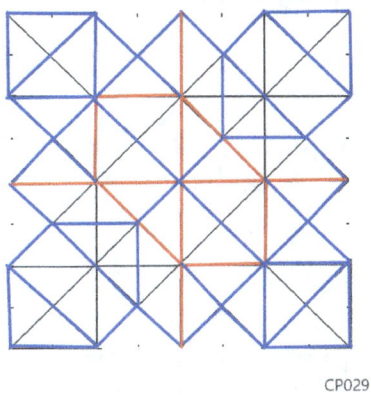

CP029

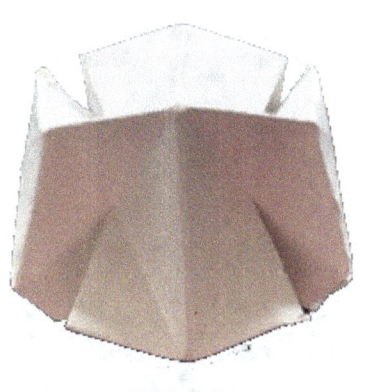

T029

T030

Model T030 has two pentagon compartment each with one pagoda in the front and two diamonds and two triangles in the back.

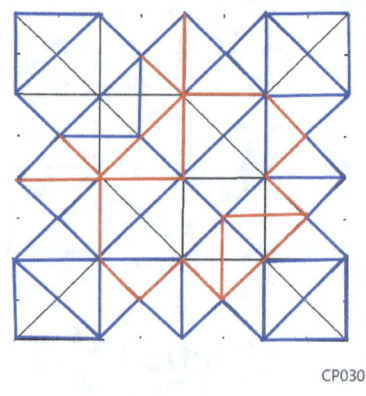

CP030

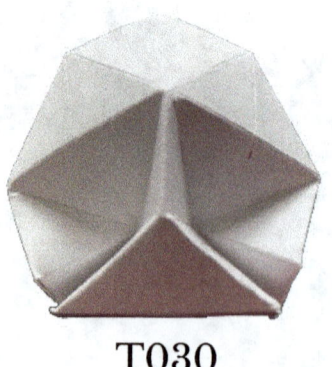

T030

T031-T045

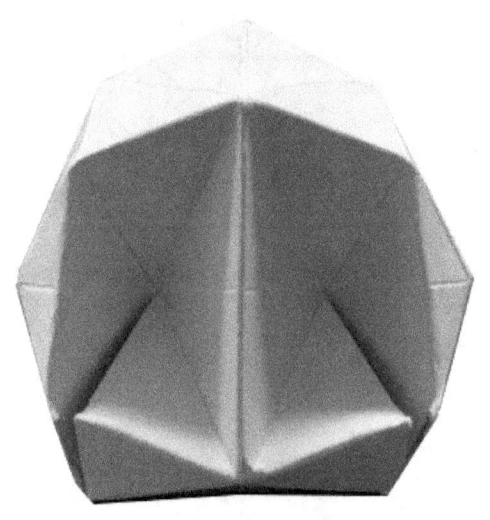

T031

Model T031 has 2 pentagon compartments each with a pagoda outward in the front and 2 triangles and 2 diamonds in the back.

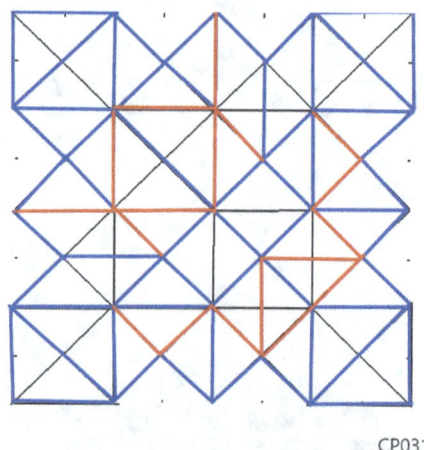

CP031

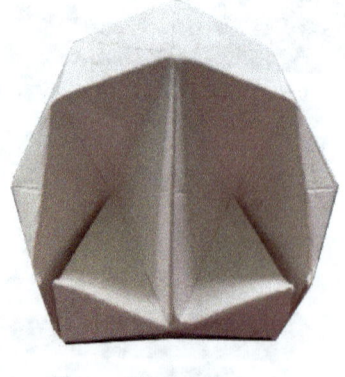

T031

T032

Model T032 has 4 pentagon compartments each with 2 pagodas inward.

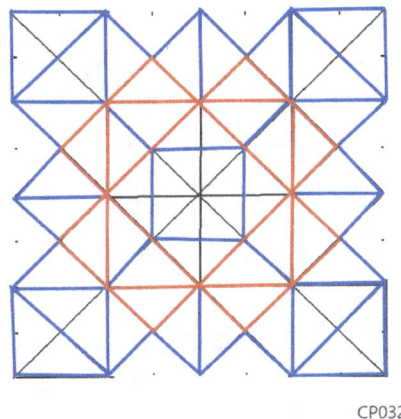

CP032

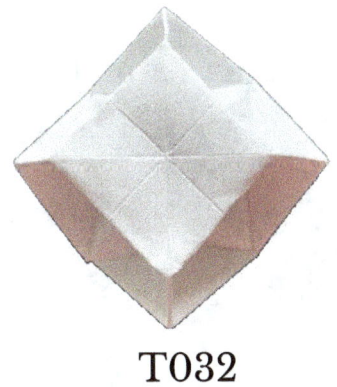

T032

T033

Model T033 has four pentagon compartments each with a pagoda inward and the other outward.

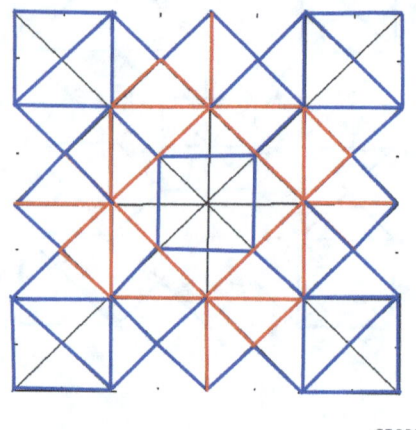

CP033

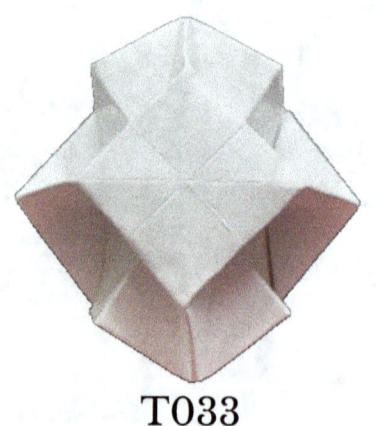

T033

T034

Model T034 has four pentagon Niches each with a pagoda outward.

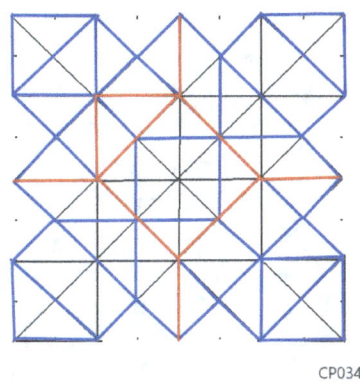

CP034

T034

T035

Model T035 has four pentagon Niches each with 2 pagodas outwards.

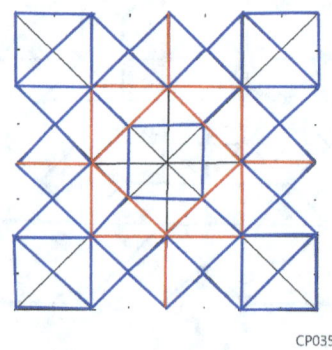

CP035

T035

T036

Model T036 has two pentagon Niches each with two pagodas inward in the front and two diamonds and two triangles in the back.

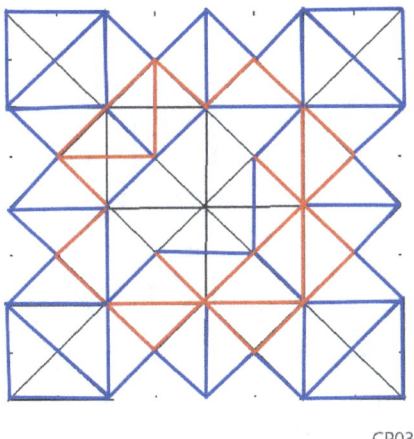

CP036

T036

T037

Model T037 has two pentagon Niches each with a pagoda outward and other one inward in the front and two diamonds and two triangles in the back.

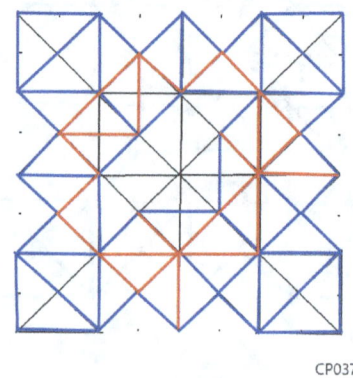

CP037

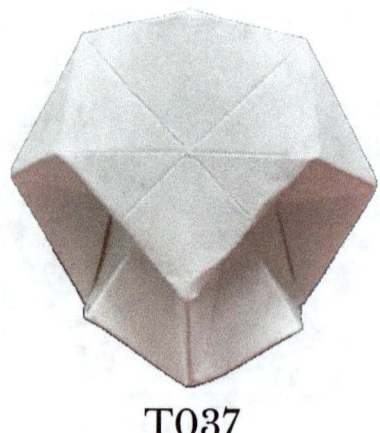

T037

T038

Model T038 has two pentagon Niches each with a pagoda outward and the other inward in the front and two diamonds and two triangles in the back.

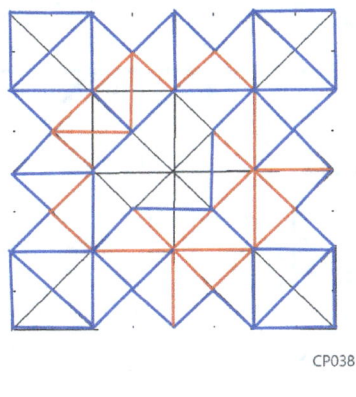

CP038

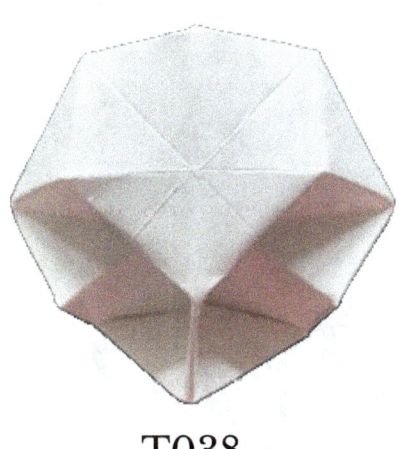

T038

T039

Model T039 has two pentagon Niches each with two pagodas outward in the front and two diamonds and two triangles in the back.

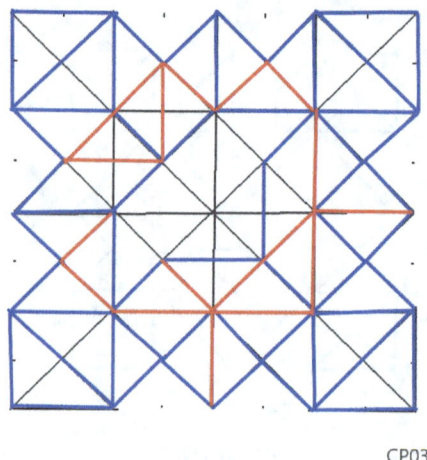

CP039

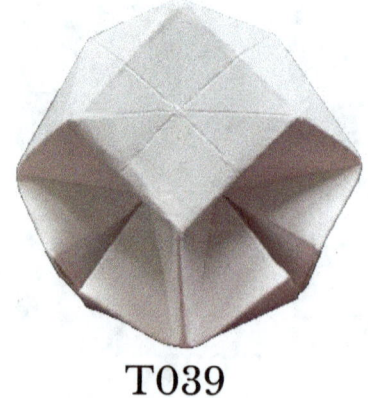

T039

T040

Model T040 has two pentagon Niches each with a pagoda outward and the other one inward in the front and two diamonds and two triangles in the back.

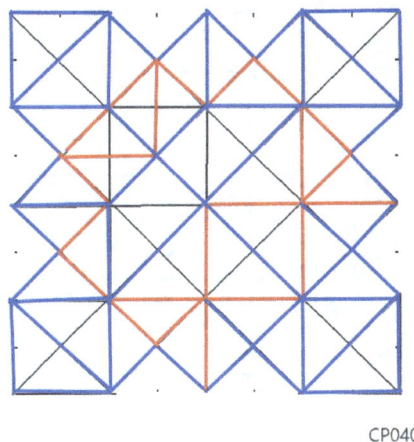

CP040

T040

T041

Model T041 has two trapezoid compartments each with one pagoda inwards. Hinge creases in double red lines applied in the model.

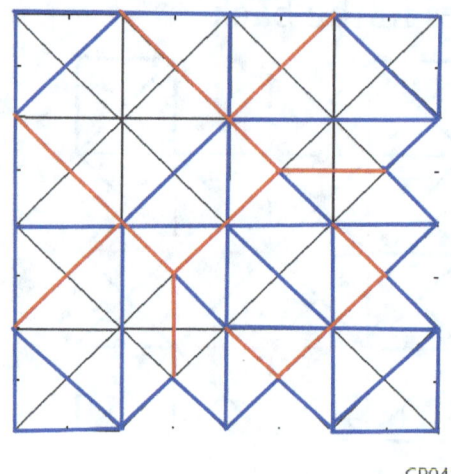

CP041

T041

T042

Model T042 has two trapezoid compartments each with one pagoda outwards. Hinge creases in double red lines applied in the model.

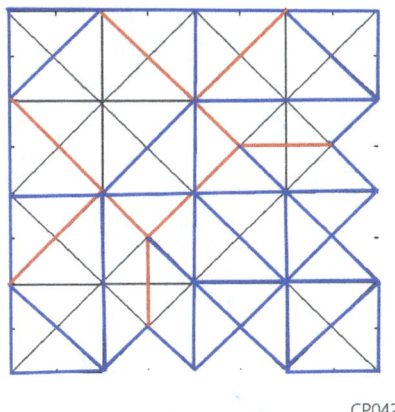

CP042

T042

T043

Model T043 has two diamonds and two trapezoids and a flat nose in the front. Hinge creases in double red lines applied in the model.

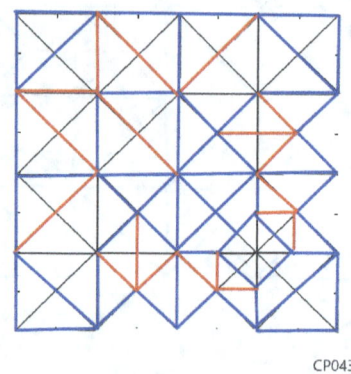

CP043

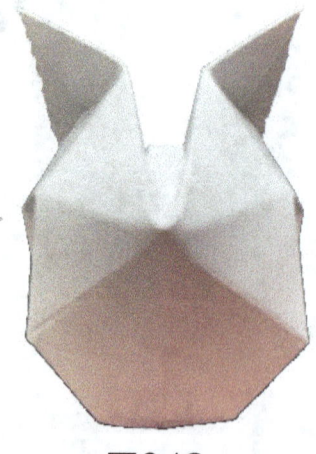

T043

T044

Model T044 has two diamonds and two triangles in the front. Hinge creases in double red lines applied in the model.

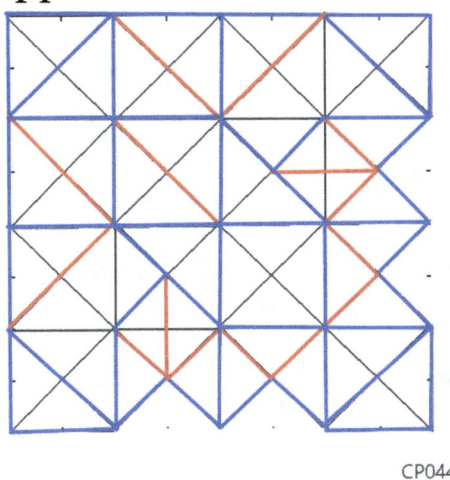

CP044

T044

T045

Model T045 has two diamonds, two trapezoids and a flat nose in the front. Hinge creases in double red lines applied in the model.

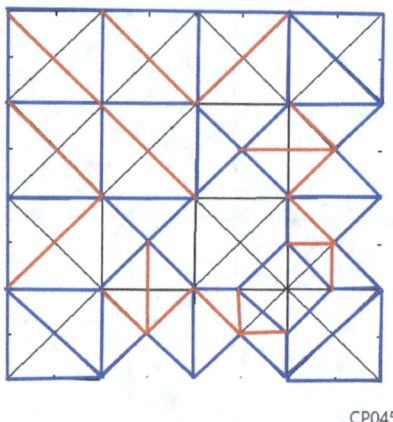

CP045

T045

T046-T060

T046

Model T046 has a large pagoda inward.

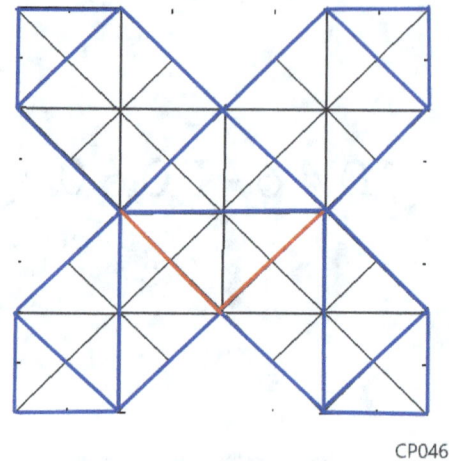

CP046

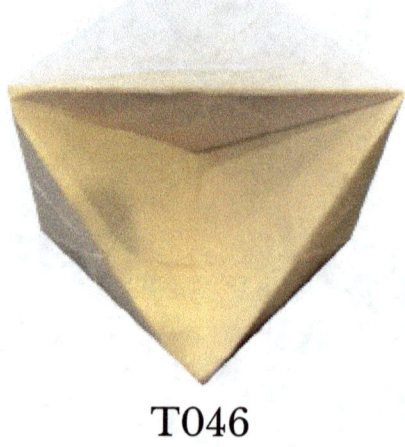

T046

T047 (1/2)

Model T047 has a large trapezoidal hole.

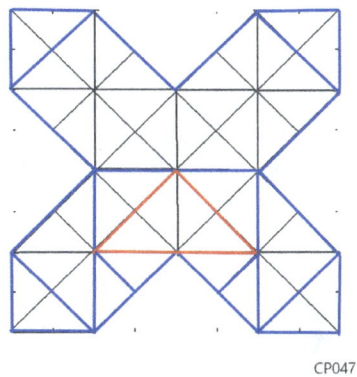

CP047

T047

T047

Masked Pig Head consists of a Mask T014, 2 x Ears T047 and Head 6 x T013.

Masked Pig Head

T048

Model T048 has two Niches in the front. Hinge creases in double red lines applied in the model.

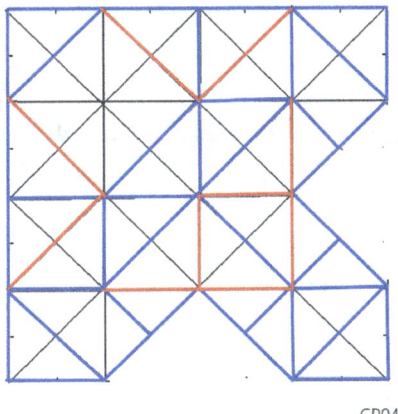

CP048

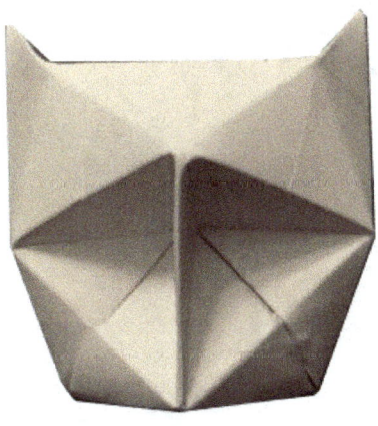

T048

T049

Model T049 has two triangles inward.

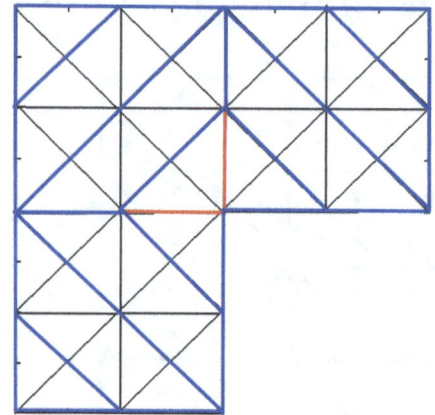

CP049

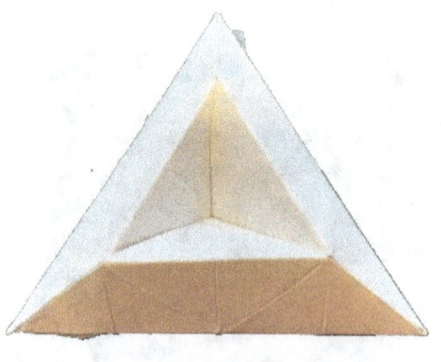

T049

T050

Model T050 has one trapezoid and one pentagon holes.

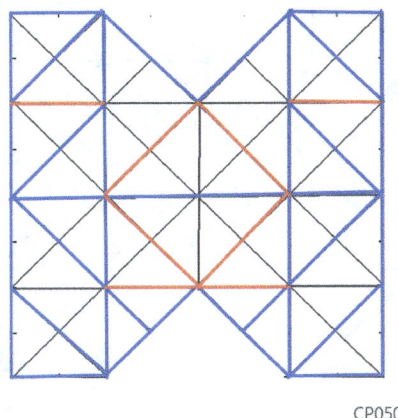

CP050

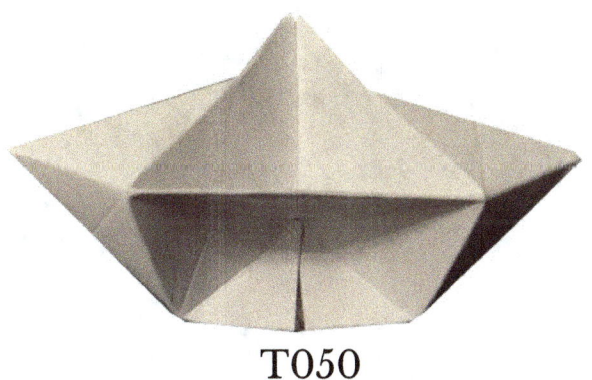

T050

T051

Model T051 has a large diamond hole.

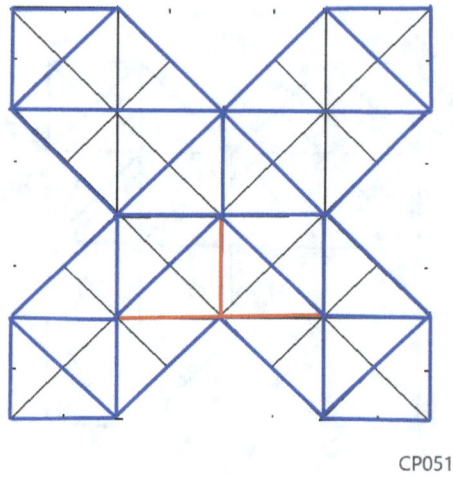

CP051

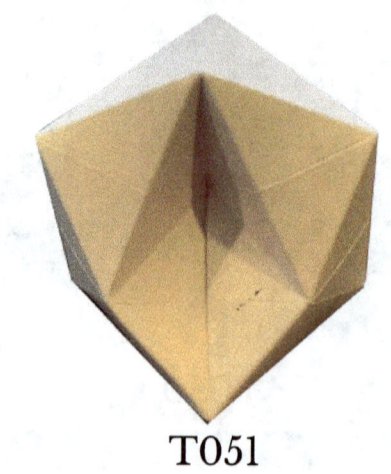

T051

T052

Model T052 has three pagoda holes.

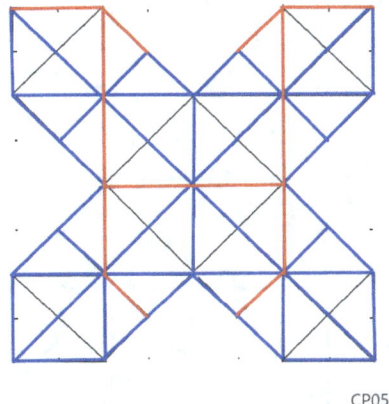

CP052

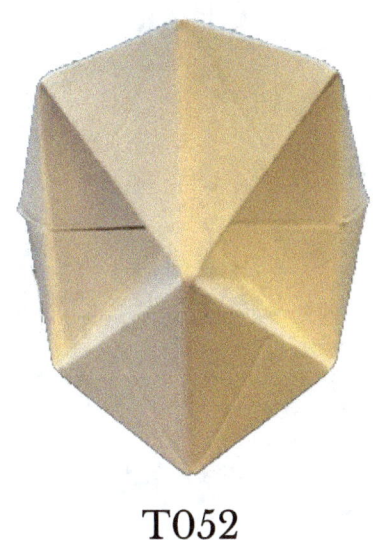

T052

T053

Model T053 has a big pagoda nose, eyes, and pagoda ears in the front and 2 triangles and 2 diamonds in the back.

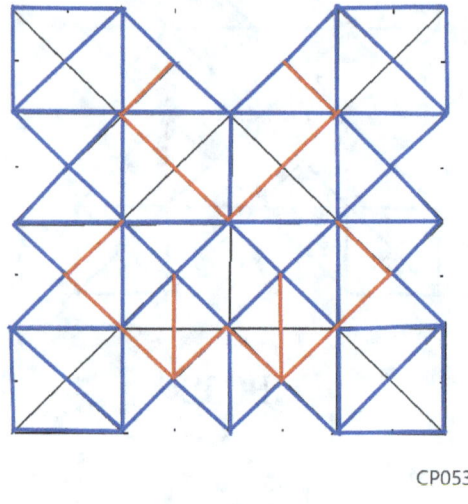

CP053

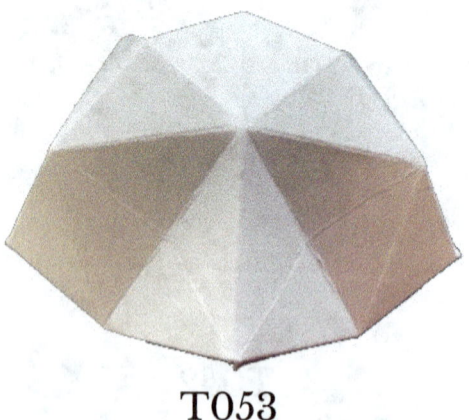

T053

T054

Model T054 has a big pagoda nose, eyes, and inward pagoda ears in the front and 2 triangles and 2 diamonds in the back.

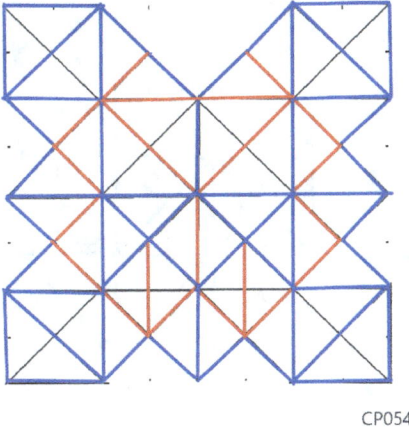

CP054

T054

T055

Model T055 has a large inward pagoda nose, eyes, and pagoda ears in the front inward.

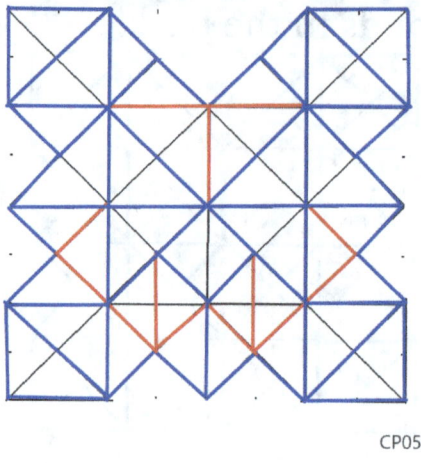

CP055

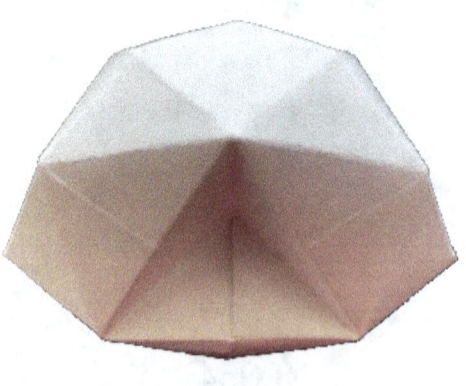

T055

T056

Model T056 has a Trapezoid hole in the front and 2 triangle and 2 diamond holes in the back.

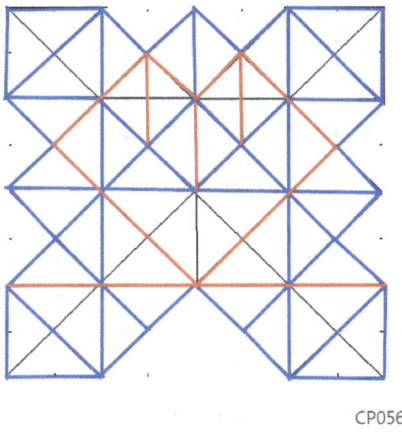

CP056

T056

T057

Model T057 has two Trapezoid holes.

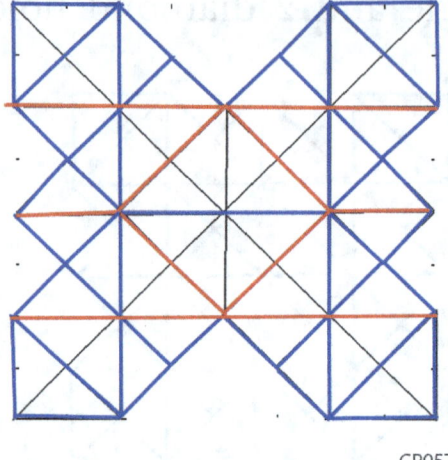

CP057

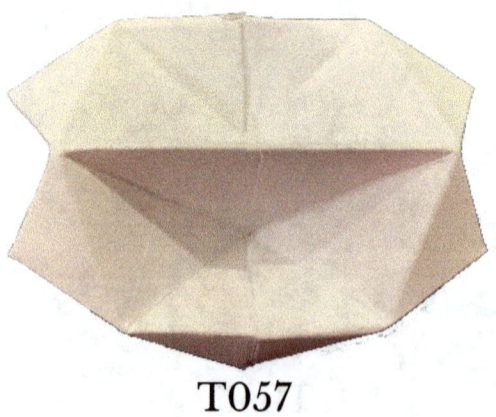

T057

T058

Model T058 has a Trapezoid hole in one side and a pagoda inward in the other.

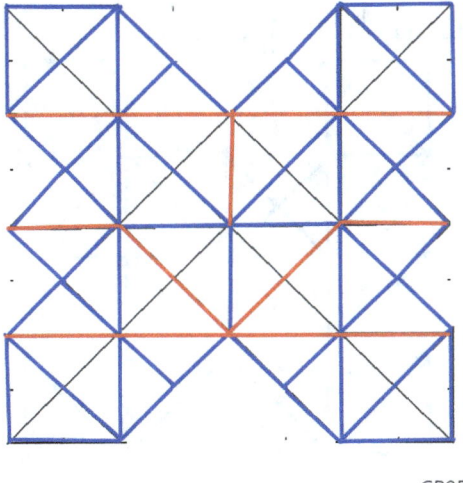

CP058

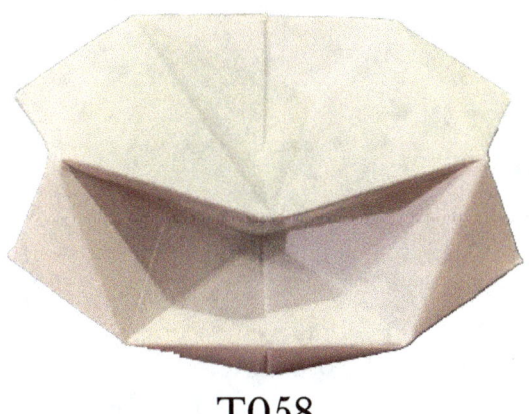

T058

T059

Model T059 has a small pagoda inside a large Trapezoid hole in one side and two diamonds and two triangle in the other side.

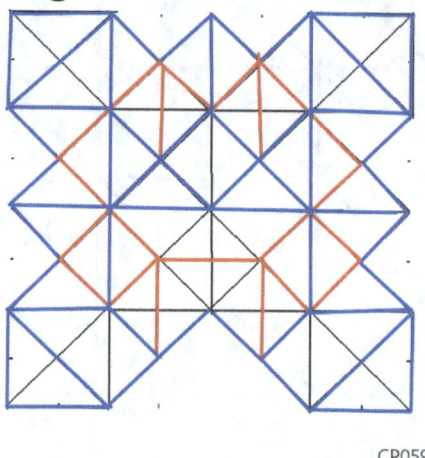

CP059

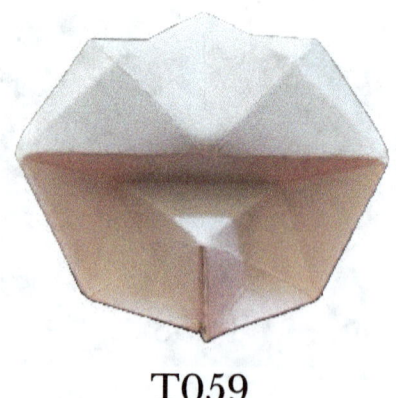

T059

T060

Model T060 has a small pagoda inside a large Trapezoid hole in the front and the same in the back.

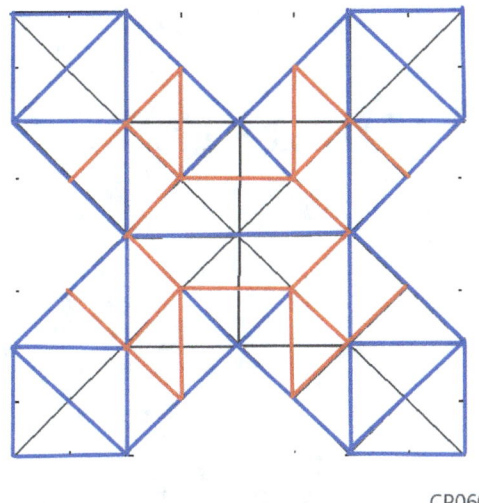

CP060

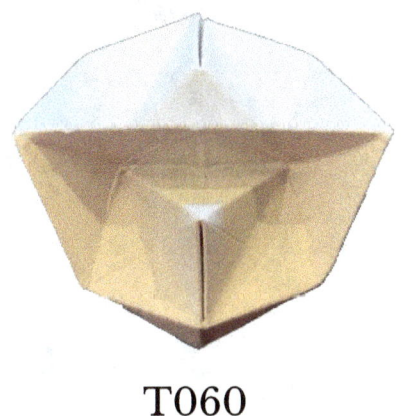

T060

T061-T075

T061

Model T061 has a big pagoda nose and two Cave crease ears.

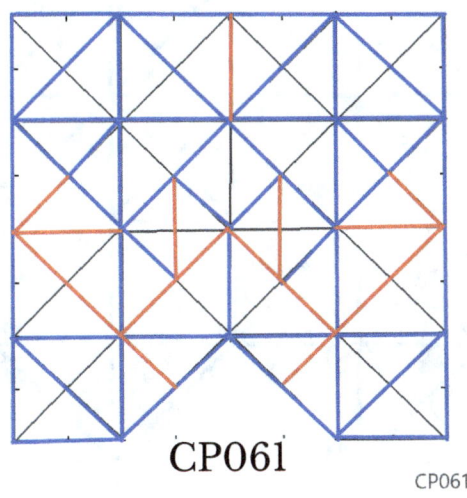

CP061

T061

T062

Model T062 has a big pagoda nose, a cone cap, and two Cave crease ears.

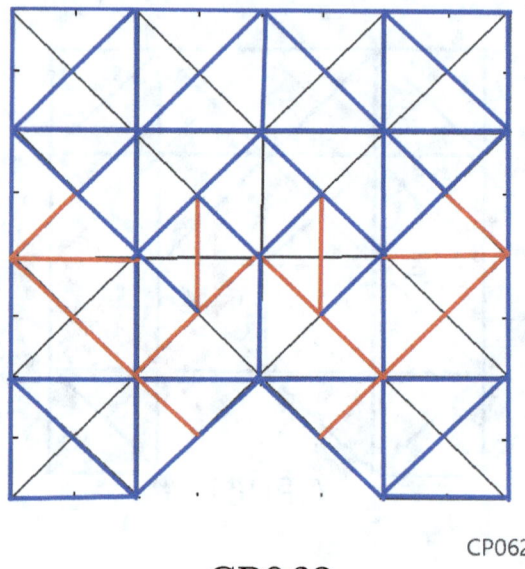

CP062

T062

T063

Model T063 has a big pagoda nose and a cone cap covered two ears.

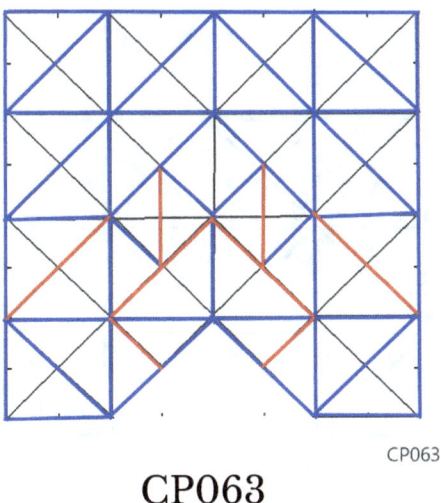

CP063

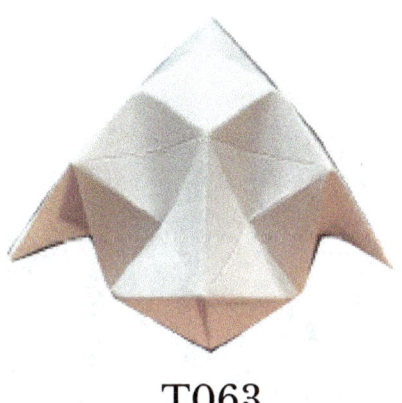

T063

T064

Model T064 has a big pagoda nose, eyes, cone cap and two big elf ears.

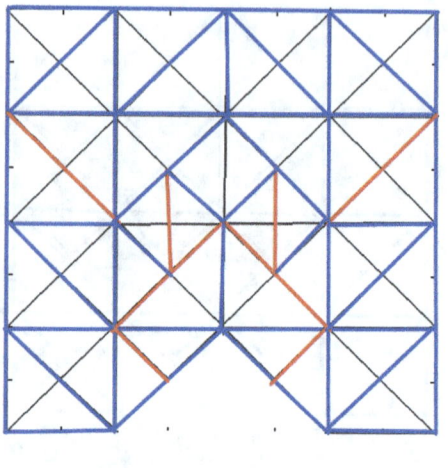

CP064

T064

T065

Model T065 has a big pagoda nose, two horns, and two Cave crease ears.

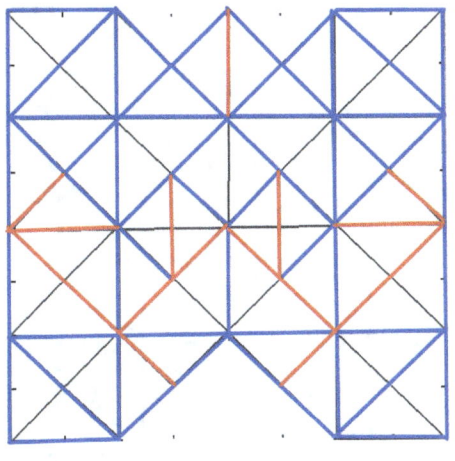

CP065

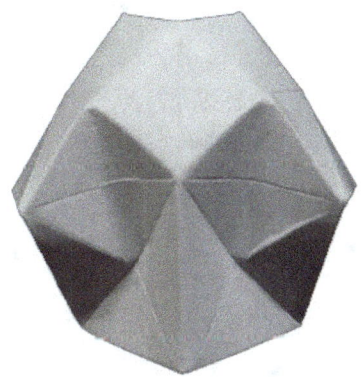

T065

T066

Model T066 has a big pagoda nose, a cap covered ears, and two horns.

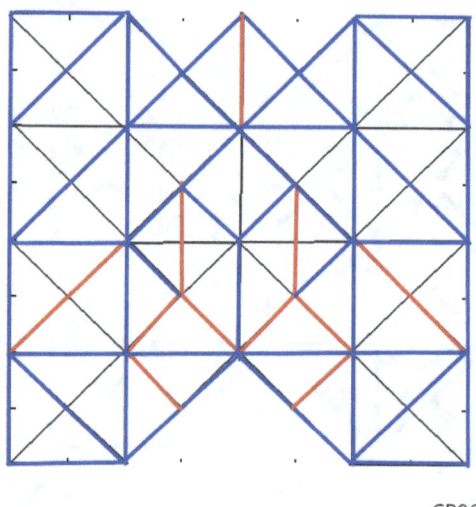

CP066

T066

T067

Model T067 has a big pagoda nose, two elf ears, and two horns.

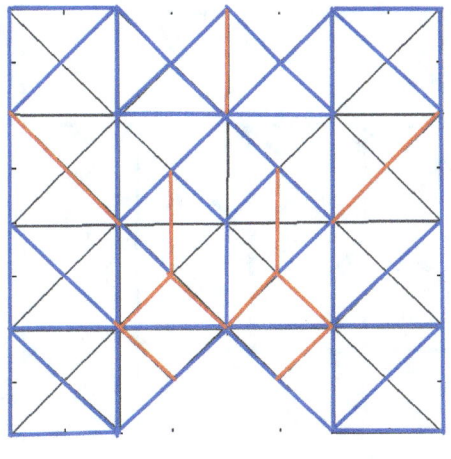

CP067

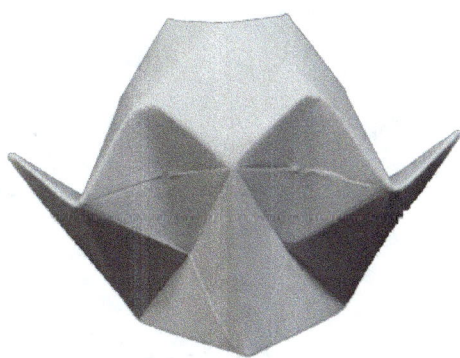

T067

T068

Model T068 has a big pagoda nose, eyes, and two Cave crease ears with long hair.

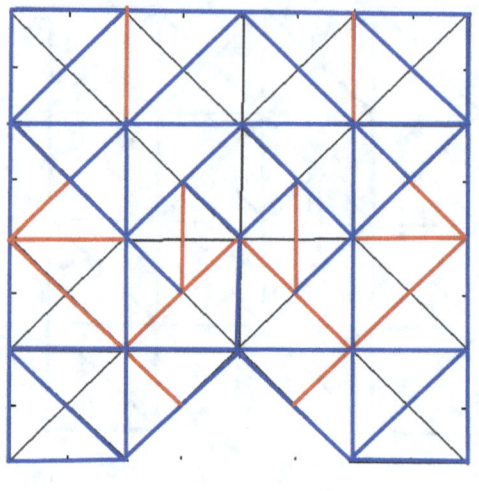

CP068

T068

T069

Model T069 has a big pagoda nose and a cap covered two ears with long hair.

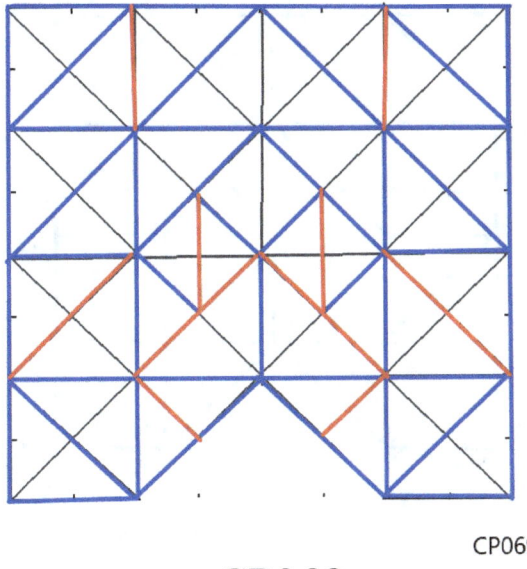

CP069

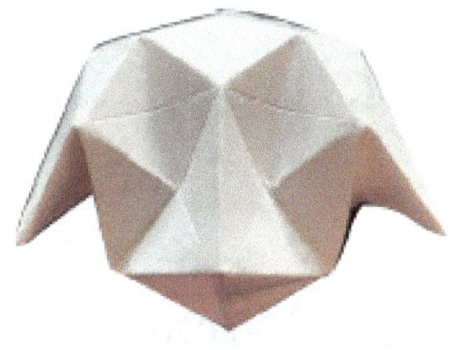

T069

T070

Model T070 has a big pagoda nose, eyes, and two elf ears with long hair.

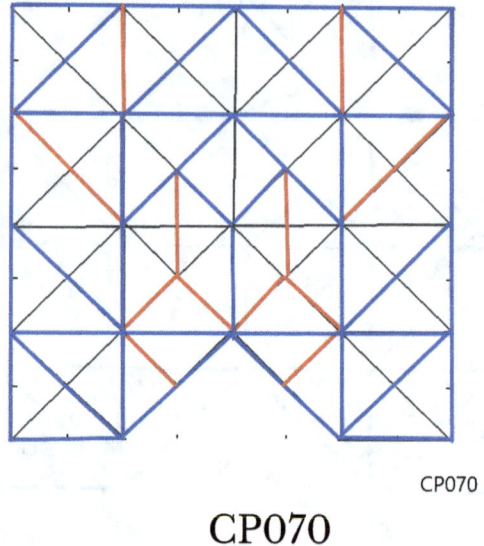

CP070

T070

T071

Model T071 has three Niches and two horns.

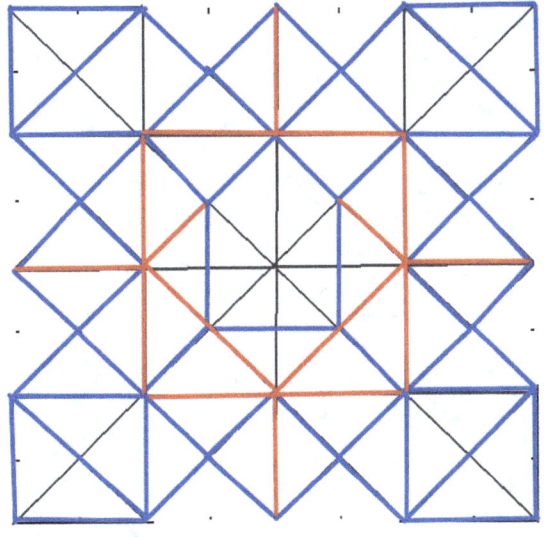

CP071

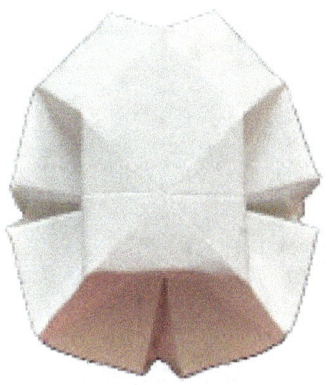

T071

T072

Model T072 has two Niche eyes, two horns, and moustache.

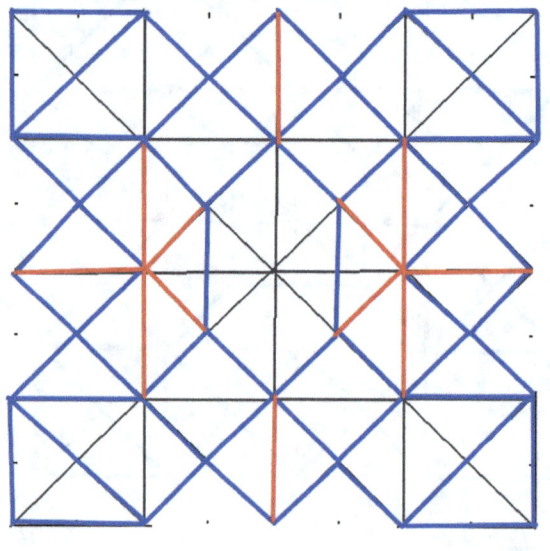

CP072

T072

T073

Model T073 has two Niche eyes, a big pagoda nose and two horns.

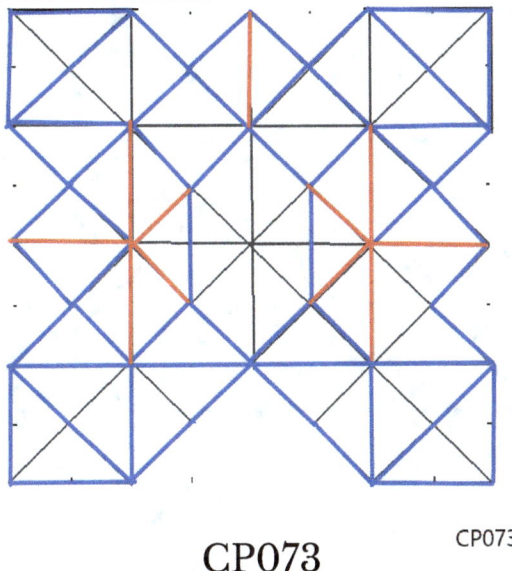

CP073

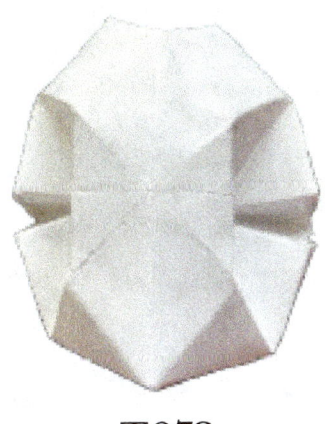

T073

T074

Model T074 has two Niche eyes and a big pagoda nose.

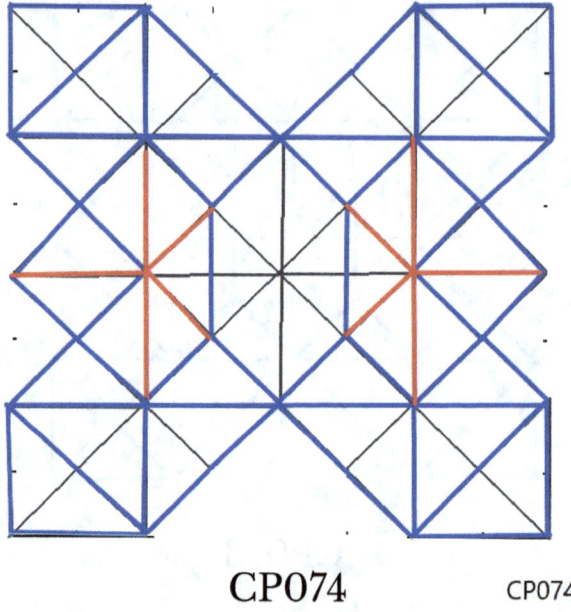

CP074   CP074

T074

T075

Model T075 has a big pagoda nose and three Niche eyes and hair.

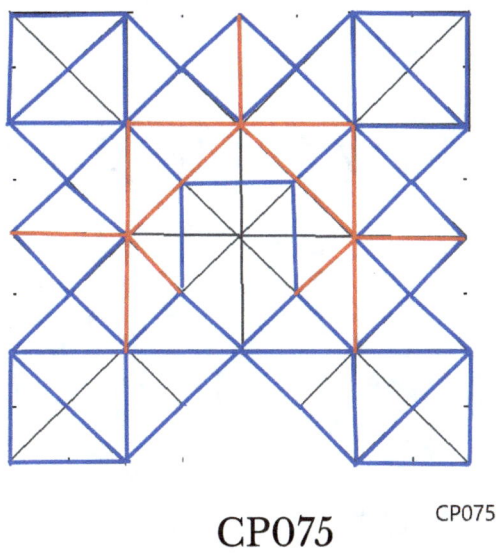

CP075

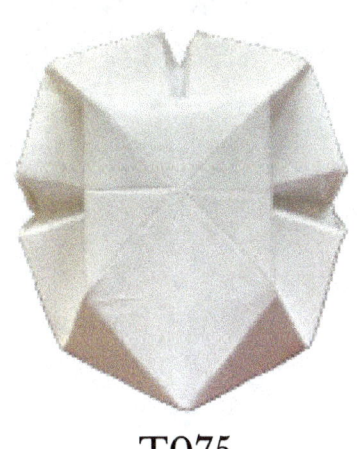

T075

T076-T090

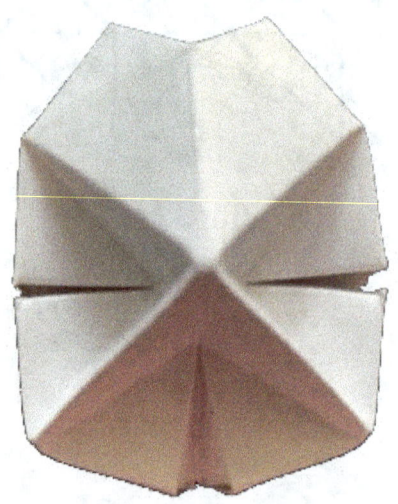

T076

Model T076 has three Niches and two horns.

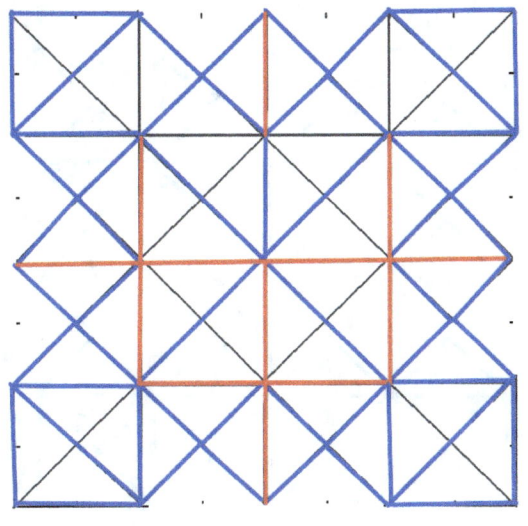
CP076

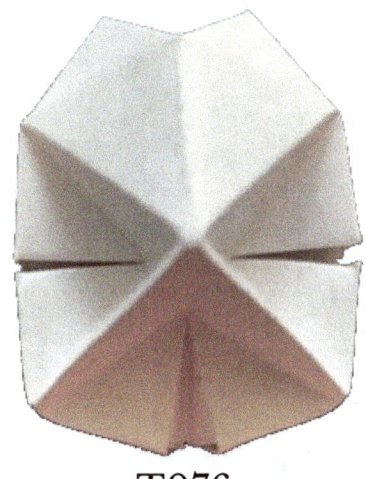
T076

T077

Model T077 has two Niches and two horns with mustache.

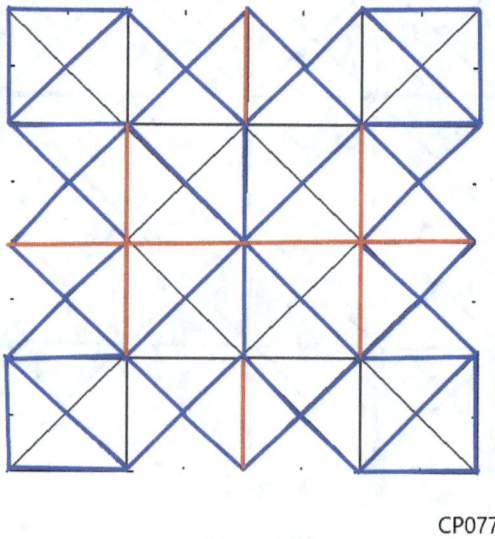

CP077

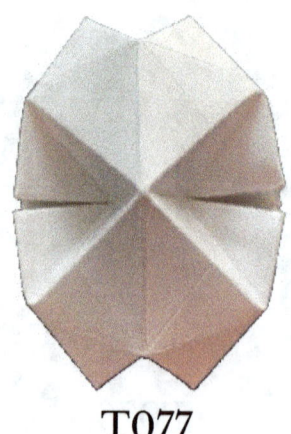

T077

T078

Model T078 has a big pagoda nose, two Niches and two horns.

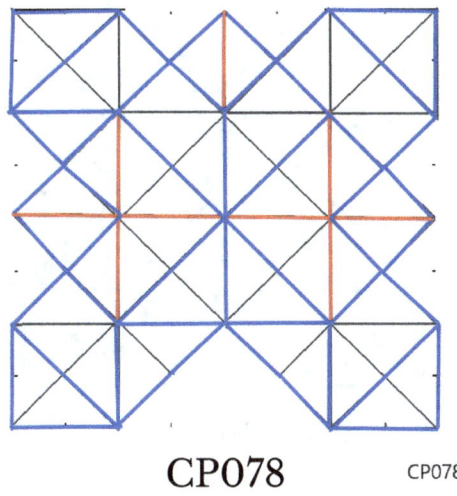

CP078 CP078

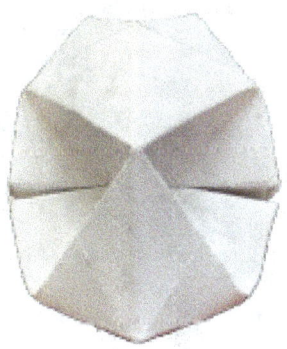

T078

T079

Model T079 has a big pagoda nose and two Niche eyes.

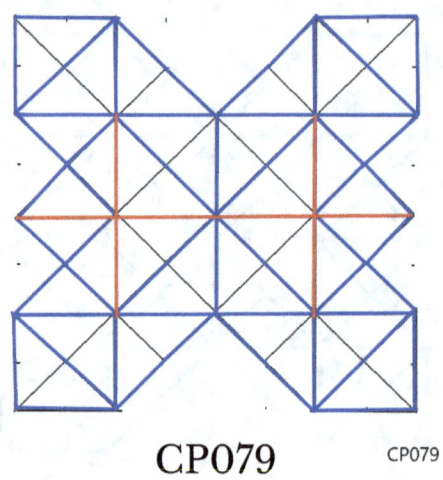

CP079

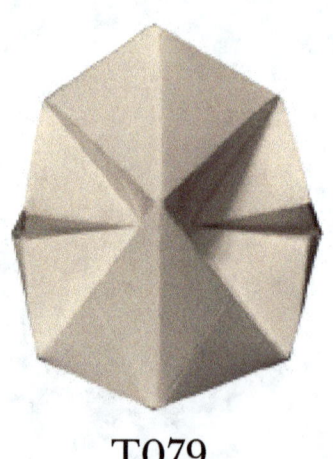

T079

T080

Model T080 has a big pagoda nose and three Niches.

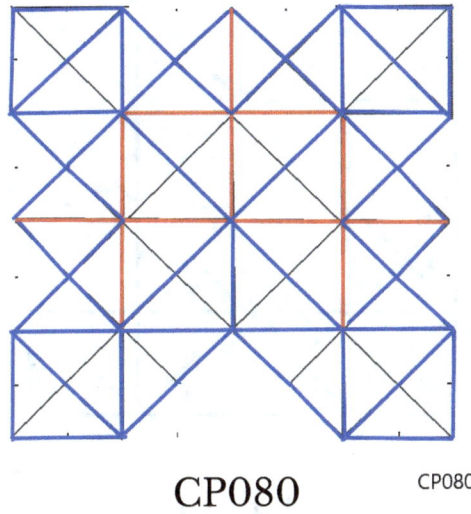

CP080

T080

T081

Model T081 has four Cave creases.

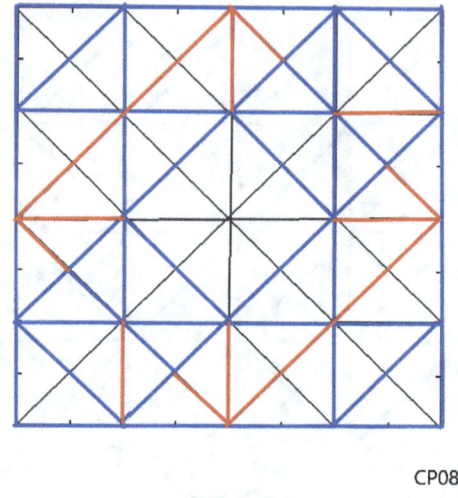

CP081

T081

T082

Model T082 has a nose with a big cone hat. A pair of Hinge crease applied.

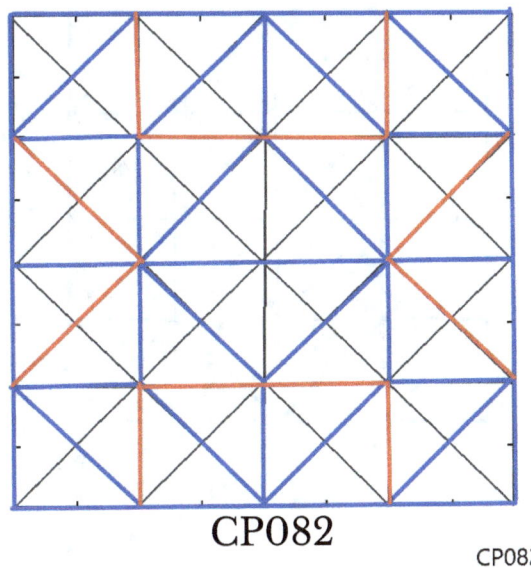

CP082

T082

T083

Model T083 has a nose with a long hair. A pair of Hinge crease applied.

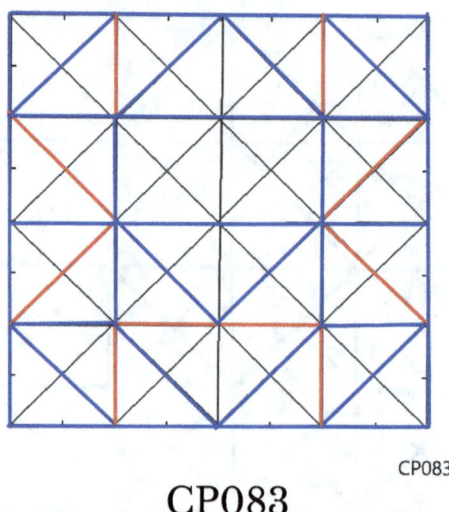

CP083

T083

T084

Model T084 has a Niche nose, two Big eyes, and two horns.

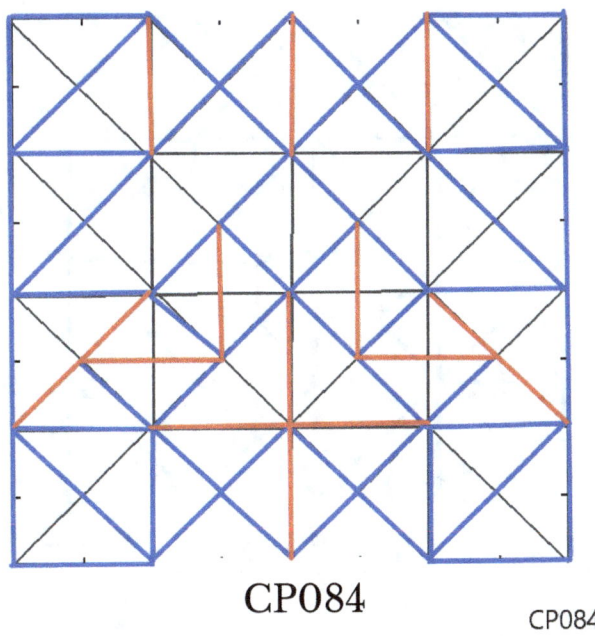

CP084

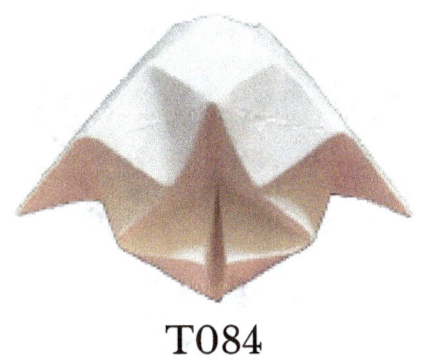

T084

T085

Model T085 has a Niche nose, two big eyes, two elf ears, and two horns.

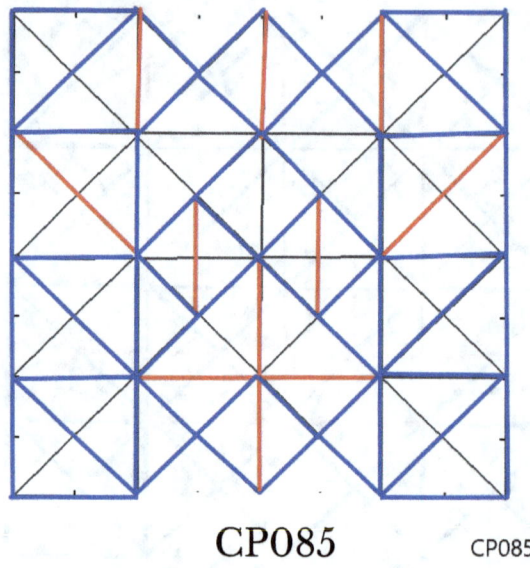

CP085 CP085

T085

T086

Model T086 has a Niche nose, two big eyes, two horns and two ears. A pair of Cave crease applied.

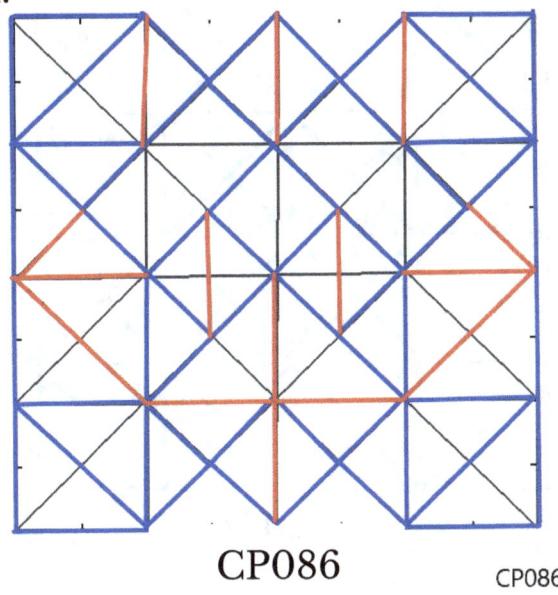

CP086

T086

T087

Model T087 has a Niche nose and two big eyes. A pair of Hinge crease applied.

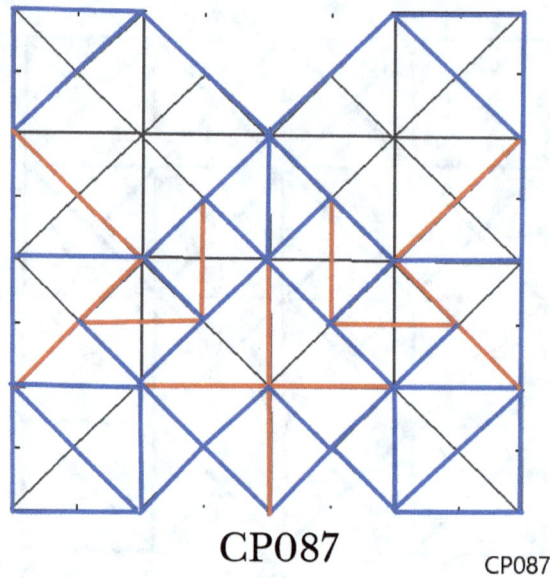

CP087

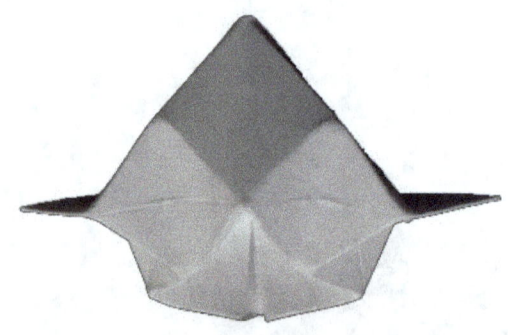

T087

T088

Model T088 has a Niche nose, and two big eyes. A pair of Hinge crease applied.

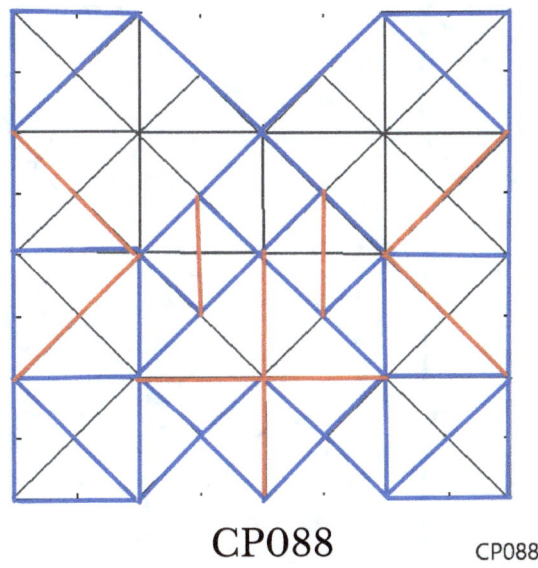

CP088   CP088

T088

T089

Model T089 has a Niche nose, two big eyes and ears. A pair of Cave crease applied.

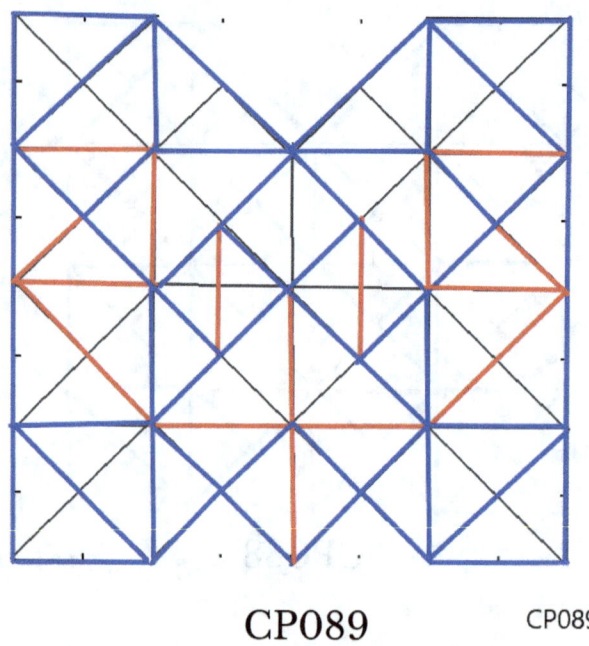

CP089   CP089

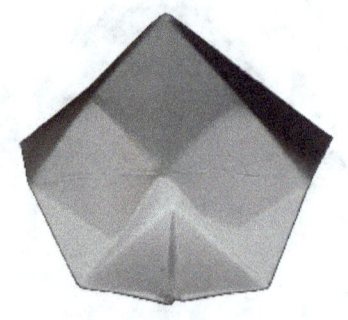

T089

T090

Model T090 is has a Niche nose, two big eyes and cone hat. A pair of Hinge crease applied.

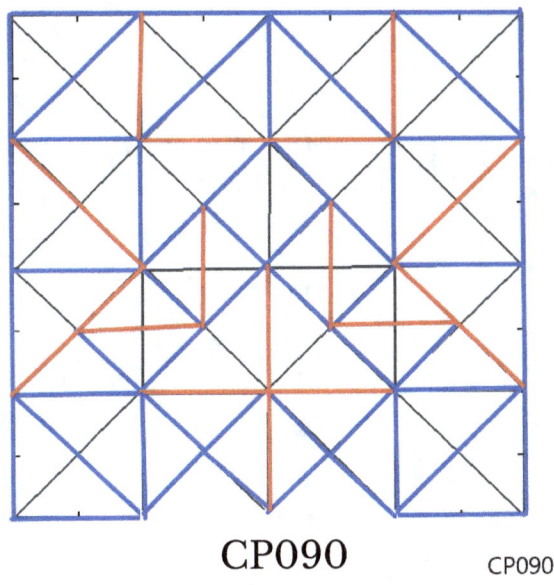

CP090 CP090

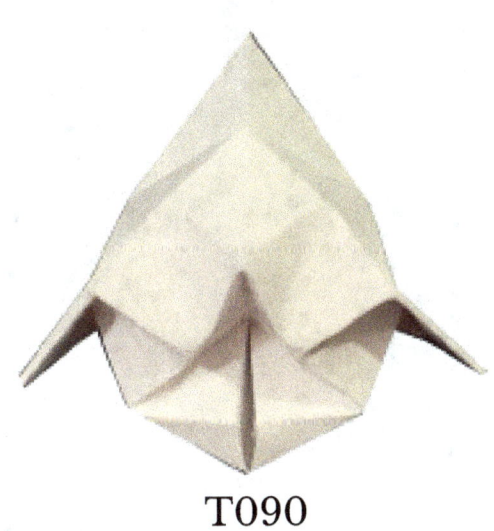

T090

T091-T105

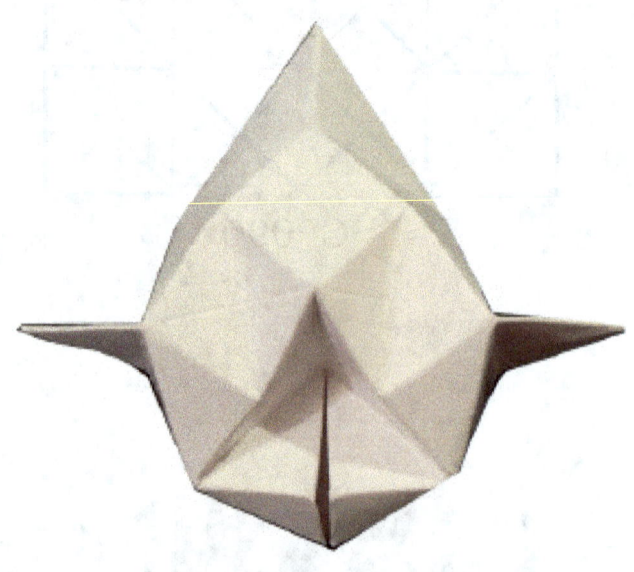

T091

Model T091 has a Niche nose, two big eyes, cone hat. A pair of Hinge crease applied.

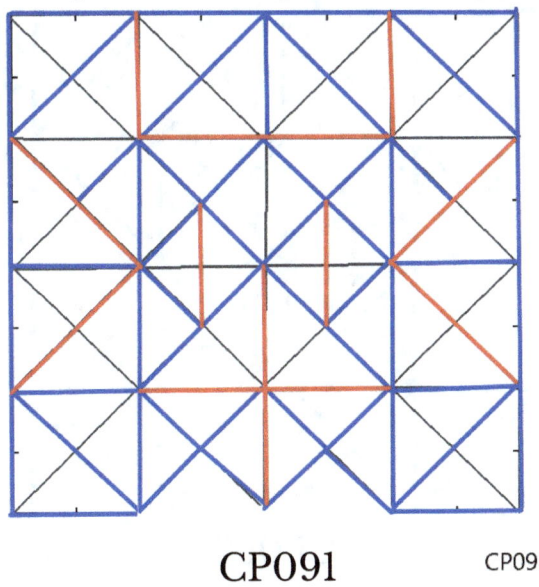

CP091   CP091

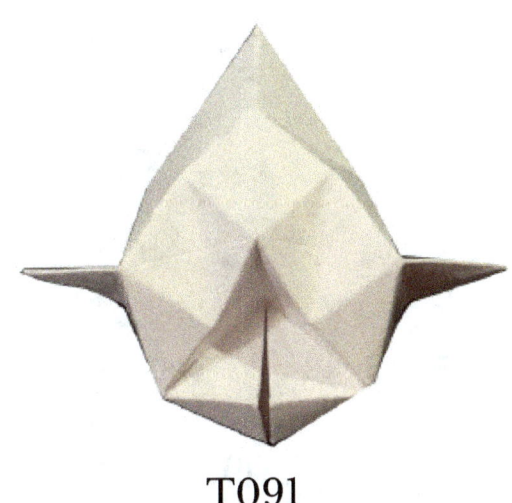

T091

T092

Model has a Niche nose, two big ears, and cone hat. A pair of Cave creases applied.

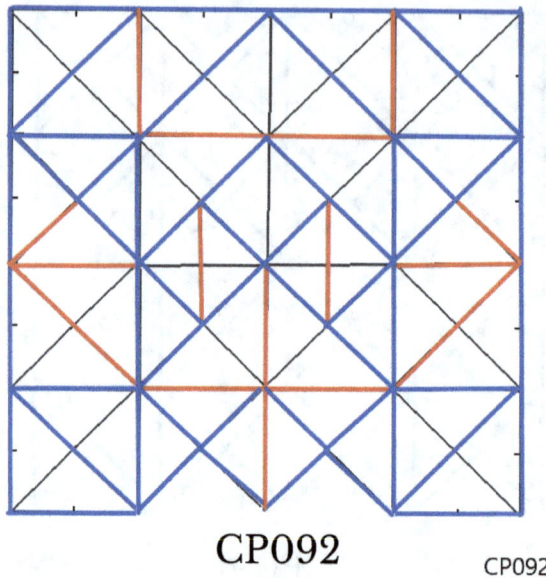

CP092

T092

T093

Model T093 has a Pagoda nose and two Niches eyes with long hair.

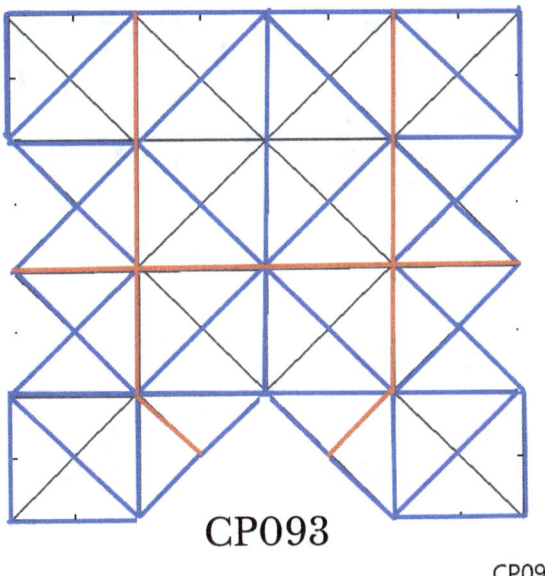

CP093

T093

T094

Model T094 has a Pagoda nose and two Niches eyes with long hair.

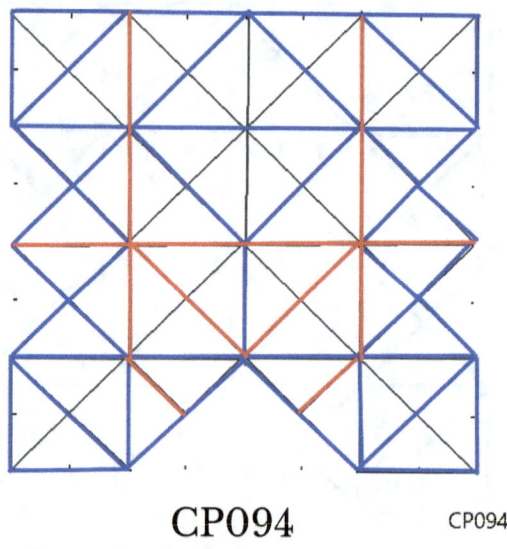

CP094

T094

T095

Model T095 a Pagoda nose, two Niches eyes and two big horns.

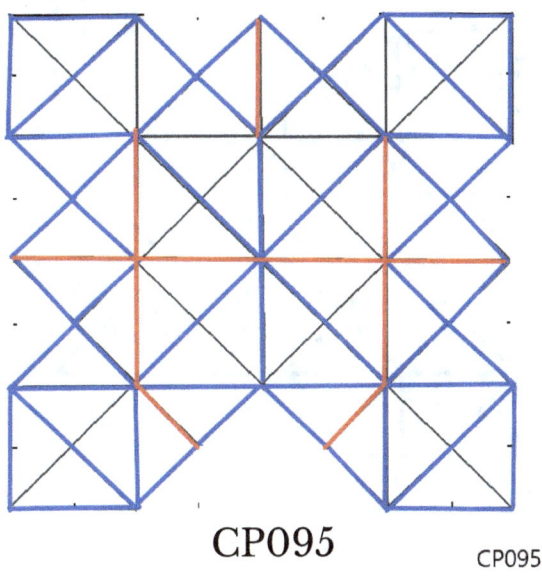

CP095

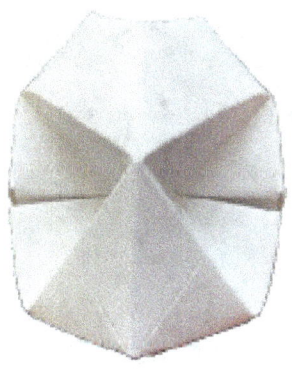

T095

T096

Model T096 a Pagoda nose, two Niches eyes and two triangles on the top.

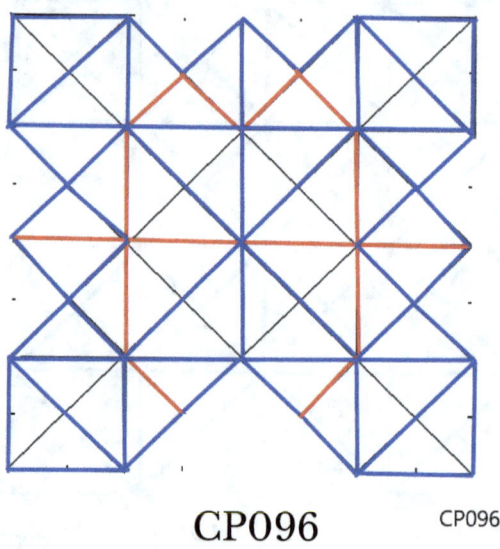

CP096 CP096

T096

T097

Model T097 has a Pagoda nose and two Niches eyes.

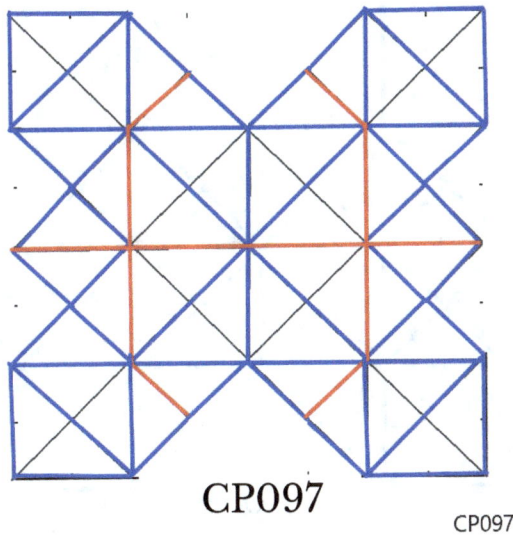

CP097

T097

T098

Model T098 has a Pagoda nose and three Niches.

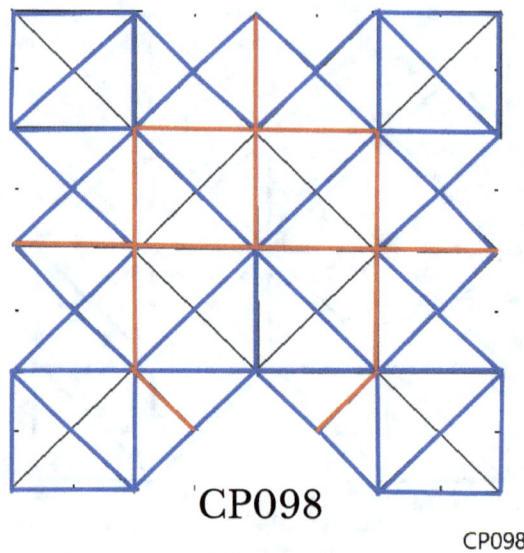

CP098

T098

T099

Model T099 has two big eyes and elf ears.

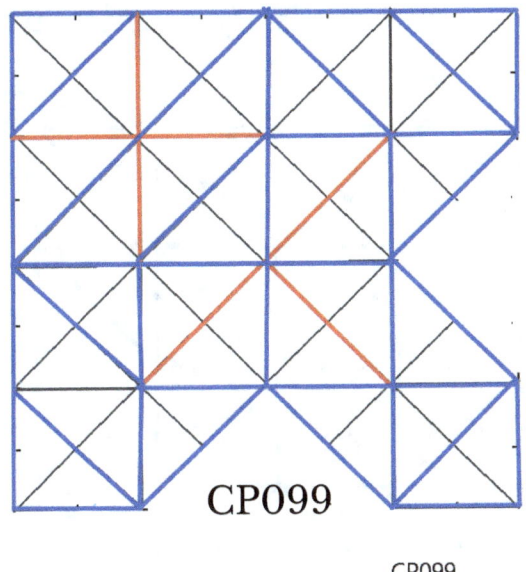

CP099

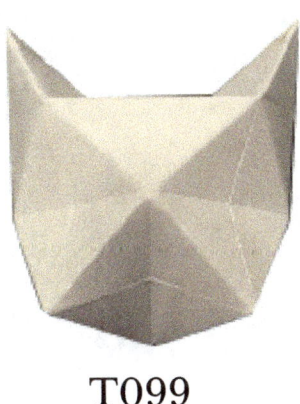

T099

T101

Model T100 has a big nose, big cave eyes, and two ears. A pair of Hinge and Cave creases applied.

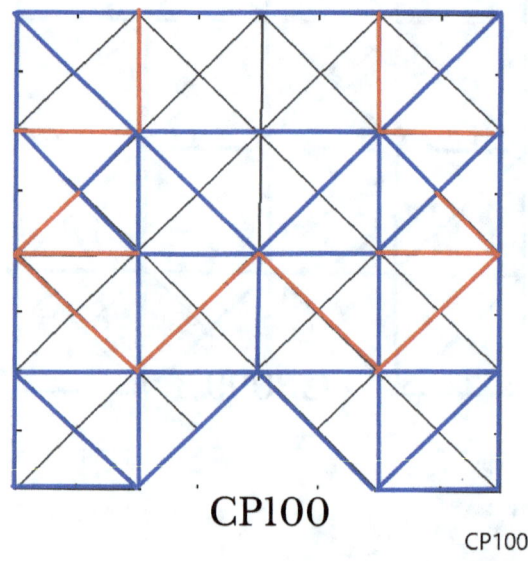

CP100

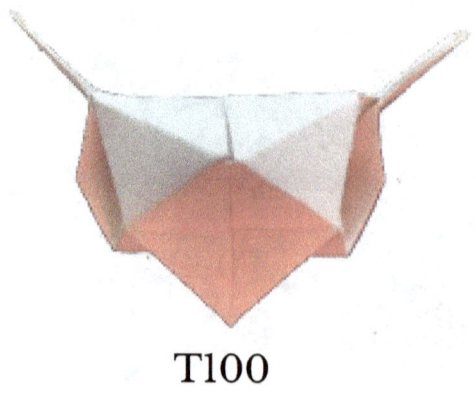

T100

T101

Model T101 has two Niche eyes and ears. A pair of Hinge creases applied.

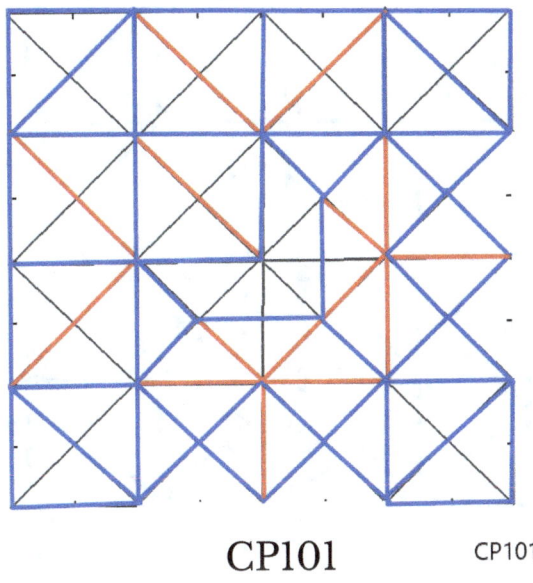

CP101 CP101

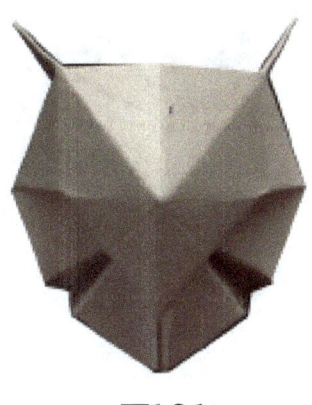

T101

T102

Model T102 has two Niche eyes and ears. A pair of Hinge creases applied.

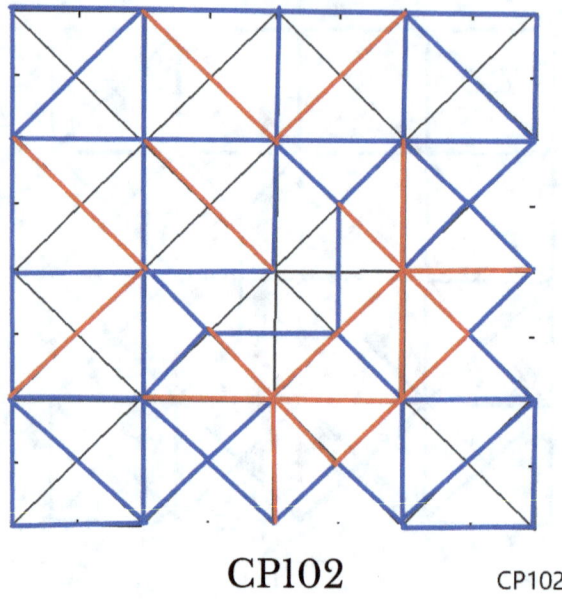

CP102 CP102

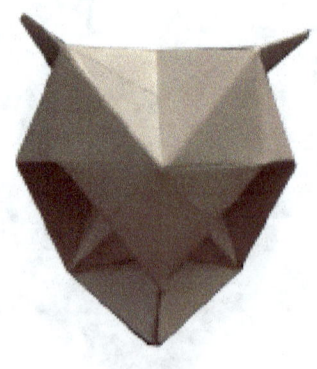

T102

T103

Model T103 has two Niche eyes and ears. A pair of Hinge creases applied.

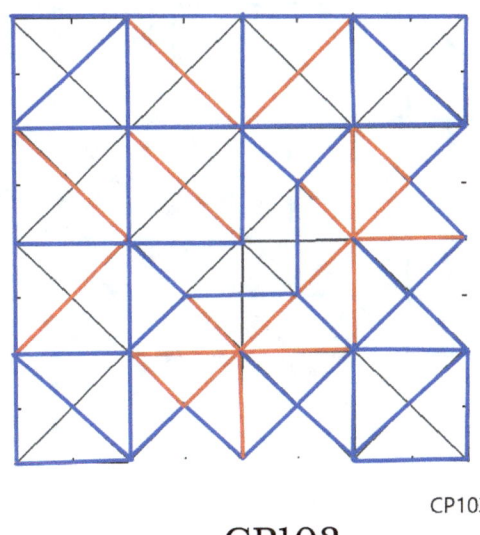

CP103

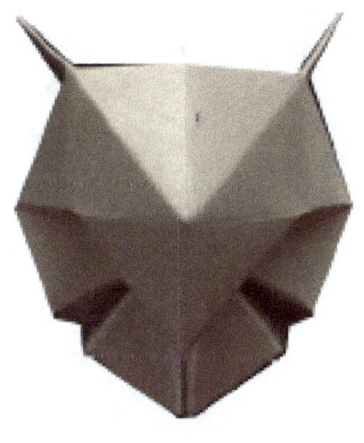

T103

T104

Model T104 has two Niche eyes and ears. A pair of Hinge creases applied.

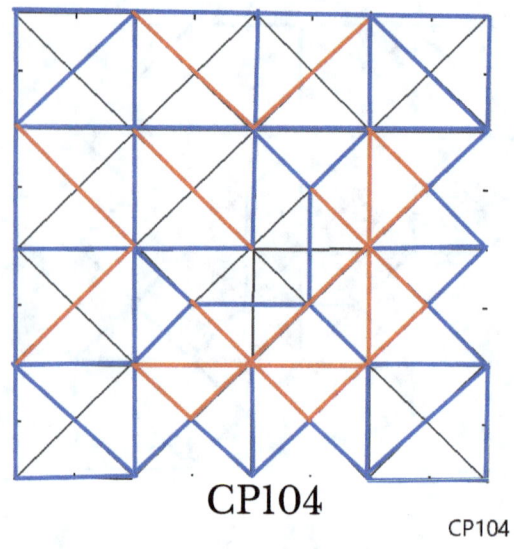

CP104

T104

T105

Model has two Niche eyes and ears. A pair of Hinge creases applied.

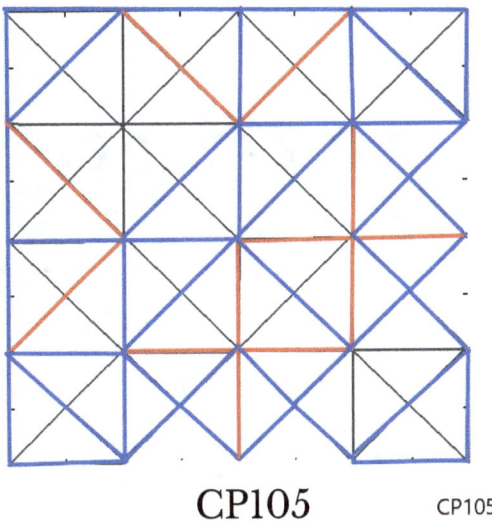

CP105 CP105

T105

T106-T120

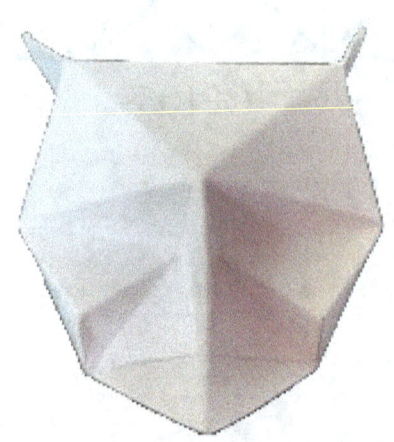

T106

Model T106 has two Niches eyes. A pair of Hinge creases applied.

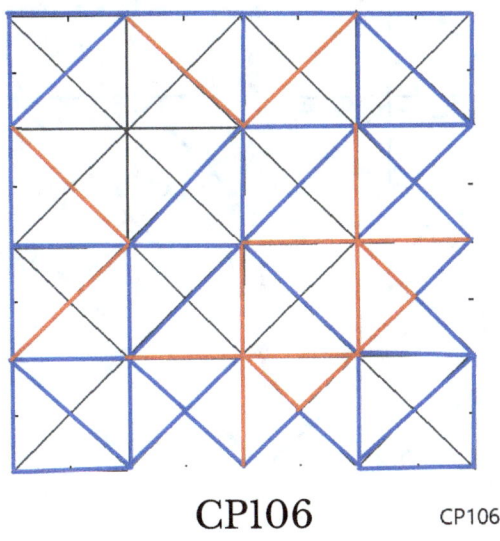

CP106

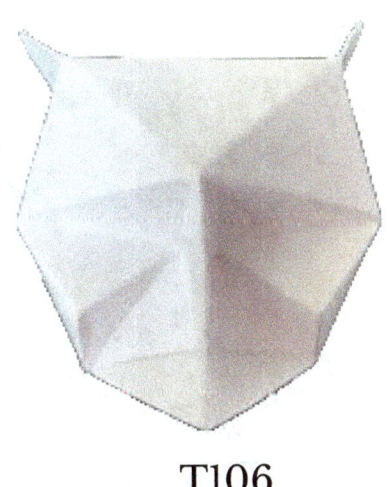

T106

T107

Model T107 has two Niches eyes. A pair of Hinge creases applied.

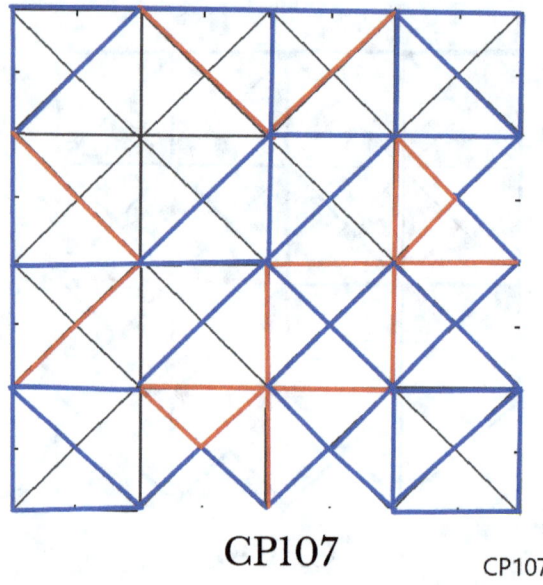

CP107 CP107

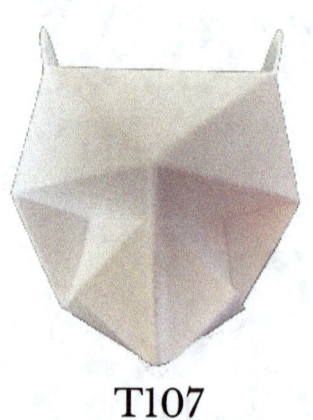

T107

T108

Model T108 has two Niches eyes. A pair of Hinge creases applied.

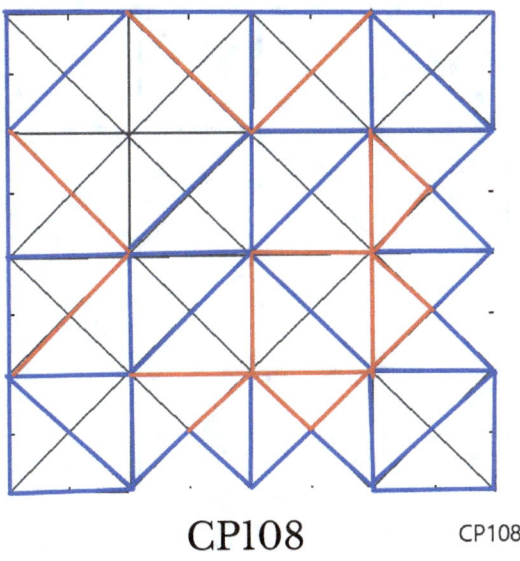

CP108   CP108

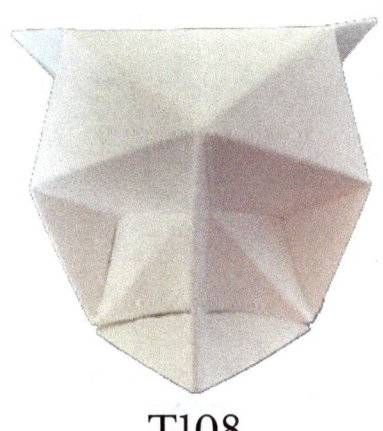

T108

T109

Model T109 has two Niches eyes. A pair of Hinge creases applied.

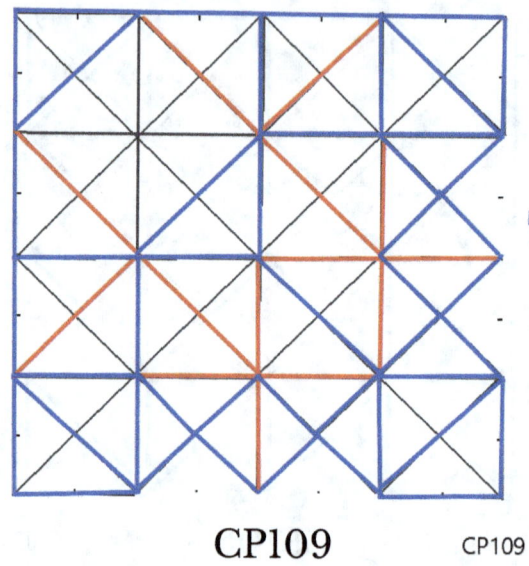

CP109 CP109

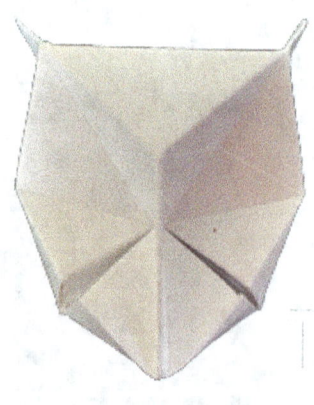

T109

T110

Model T110 has two large Niches eyes. A pair of Hinge creases applied.

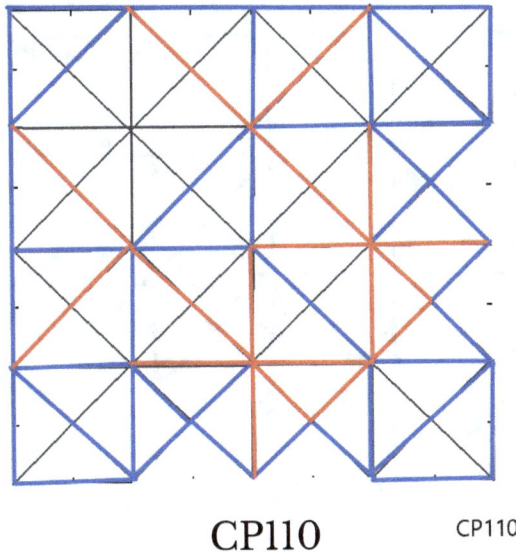

CP110   CP110

T110

T111

Model T111 has two large Niches eyes. A pair of Hinge creases applied.

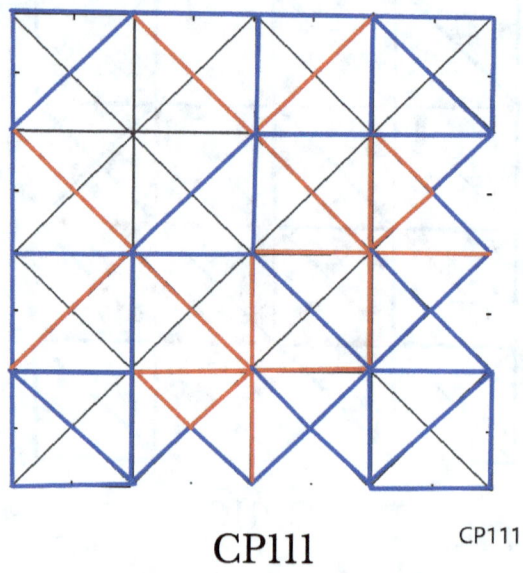

CP111 CP111

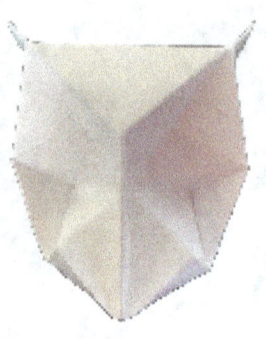

T111

T112

Model T112 has two large Niches eyes. A pair of Hinge creases applied.

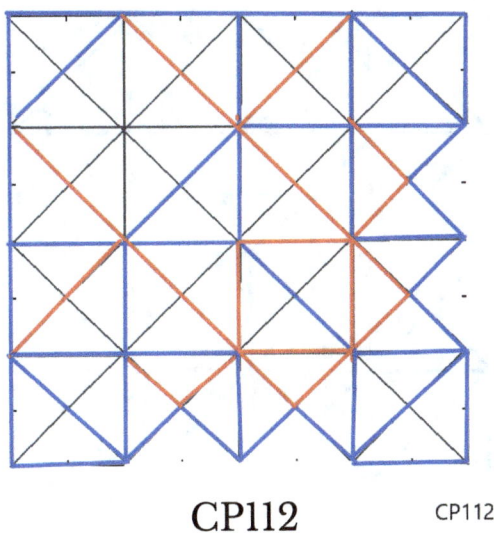

CP112   CP112

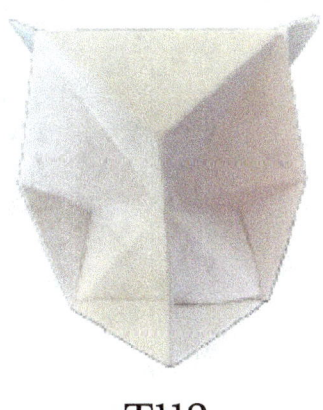

T112

T113

Model T113 has two Niches eyes in front side and a diamond and triangle set in the back side.

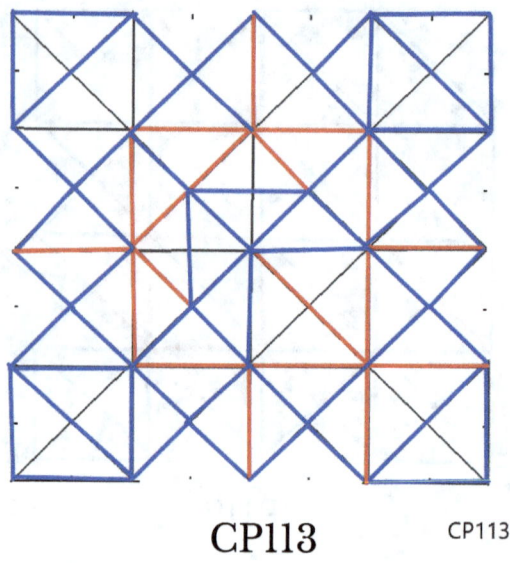

CP113   CP113

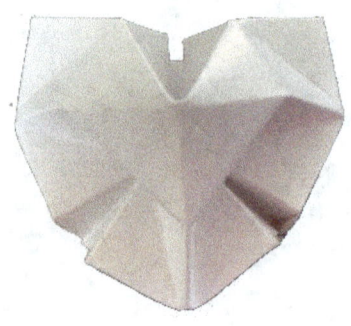

T113

T114

Model T114 has two Niches eyes in front side and a diamond and triangle set in the back side.

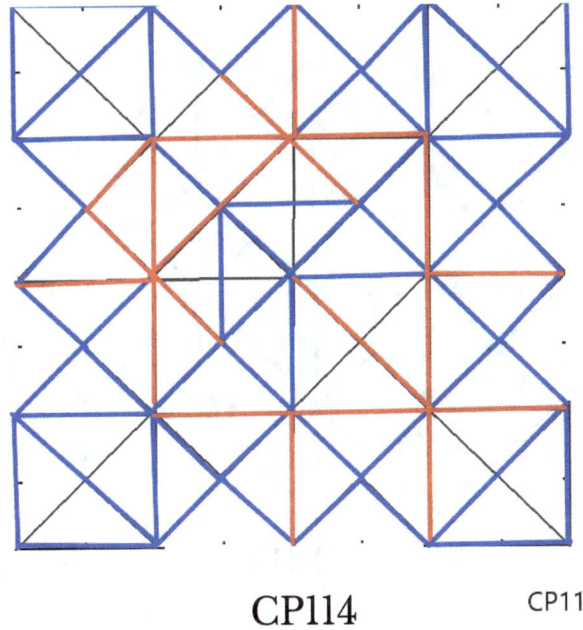

CP114   CP114

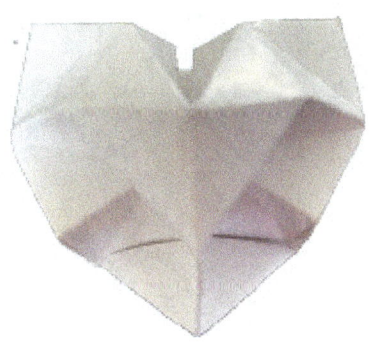

T114

T115

Model T115 has two Niches eyes in front side and a diamond and triangle set in the back side.

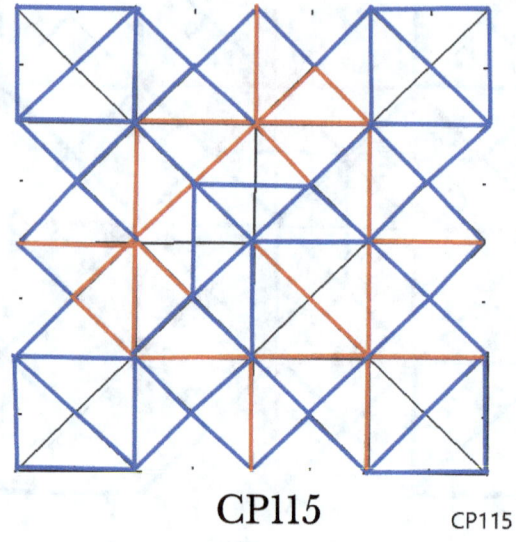

CP115 CP115

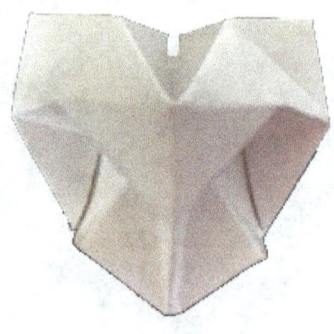

T115

T116

Model T116 has two Niches eyes in front side and a diamond and triangle set in the back side.

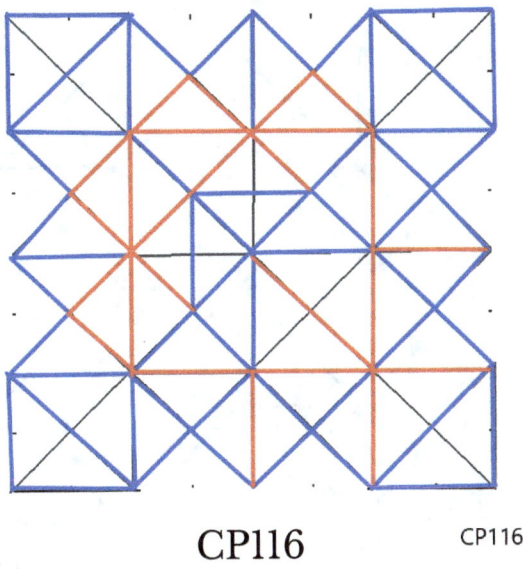

CP116   CP116

T116

T117

Model T117 has two Niches eyes in front side and a diamond and triangle set in the back side.

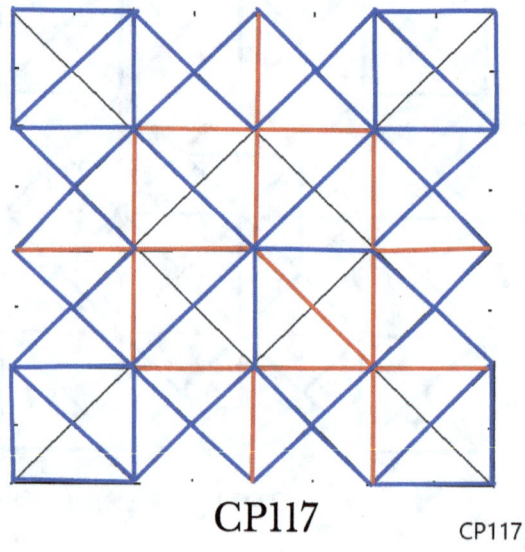

CP117   CP117

T117

T118

Model T118 has two Niches eyes in front side and a diamond and triangle set in the back side.

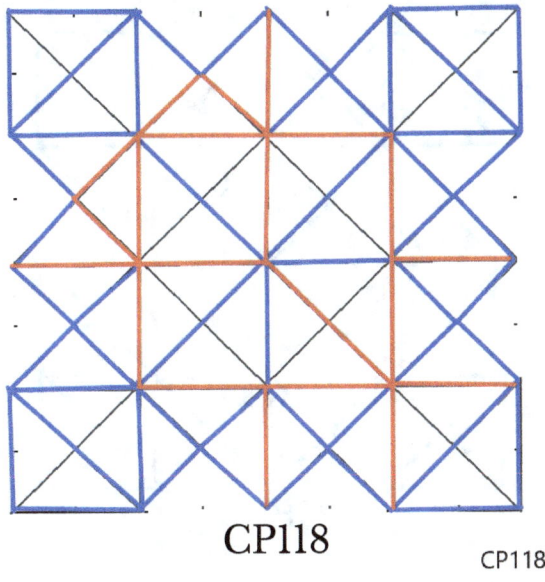

CP118

CP118

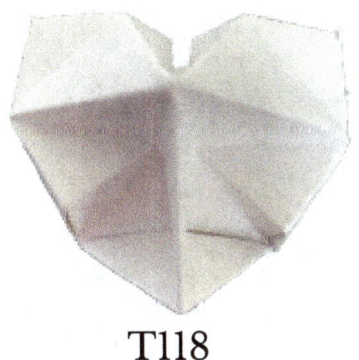

T118

T119

Model T119 is has two Niches eyes in front side and a diamond and triangle set in the back side.

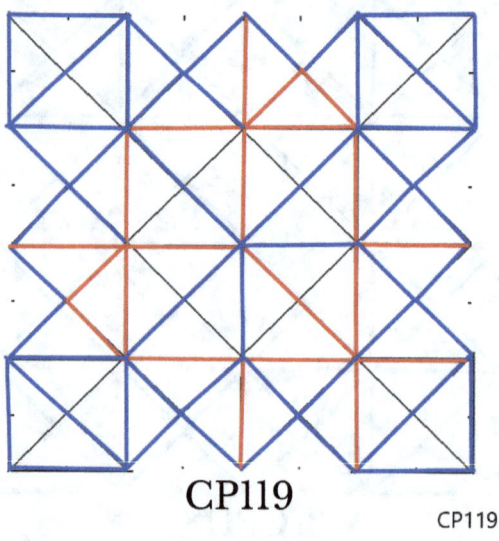

CP119

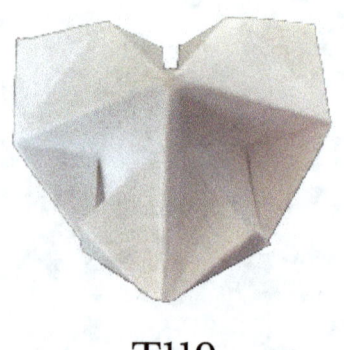

T119

T120

Model T120 has two Niches eyes in front side and a diamond and triangle set in the back side.

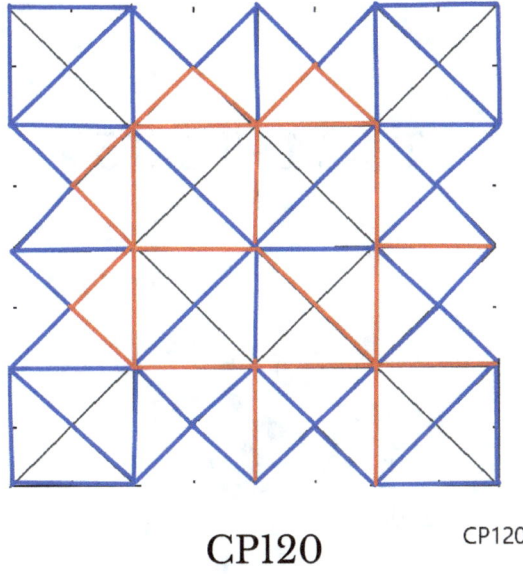

CP120   CP120

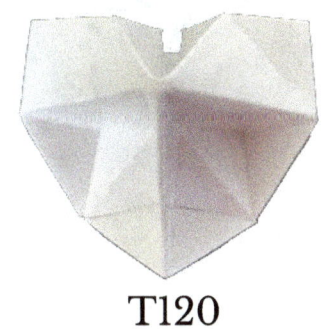

T120

T121-T135

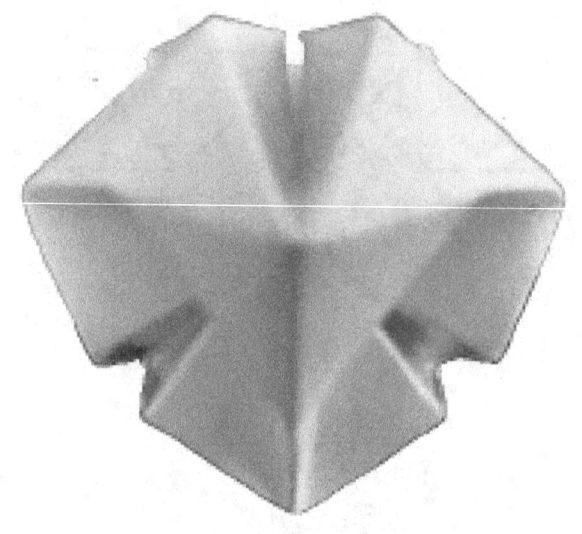

T121

Model T121 has two Niches eyes in front side and a diamond and triangle set and two triangles in the back side.

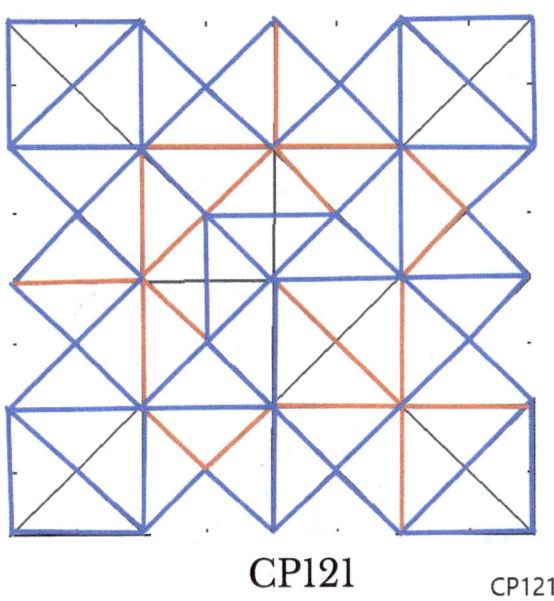

CP121   CP121

T121

T122

Model T122 has two Niches eyes in front side and a diamond and triangle set and two triangles in the back side.

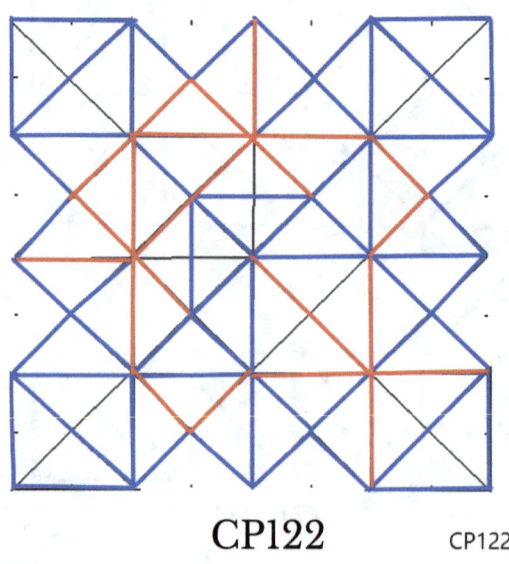

CP122   CP122

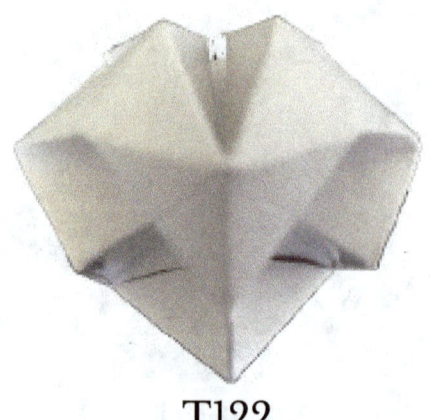

T122

T123

Model T123 has two Niches eyes in front side and a diamond and two triangle set and two triangles in the back side.

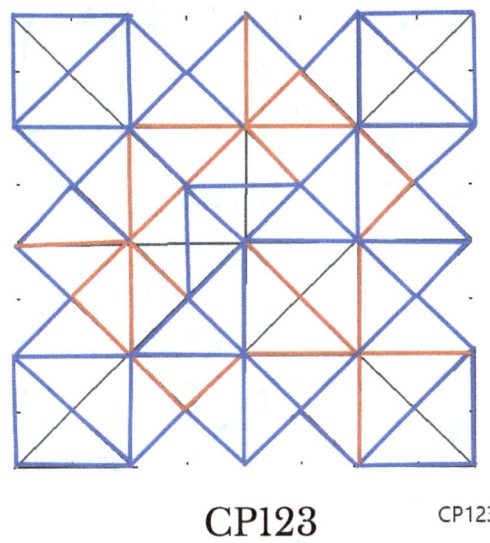

CP123   CP123

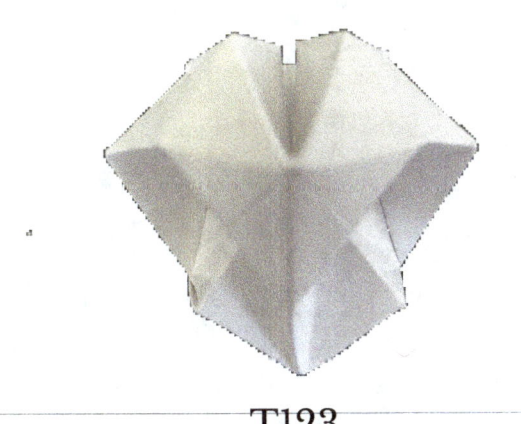

T123

T124

Model T124 has two Niches eyes in front side and a diamond and two triangle set and two triangles in the back side.

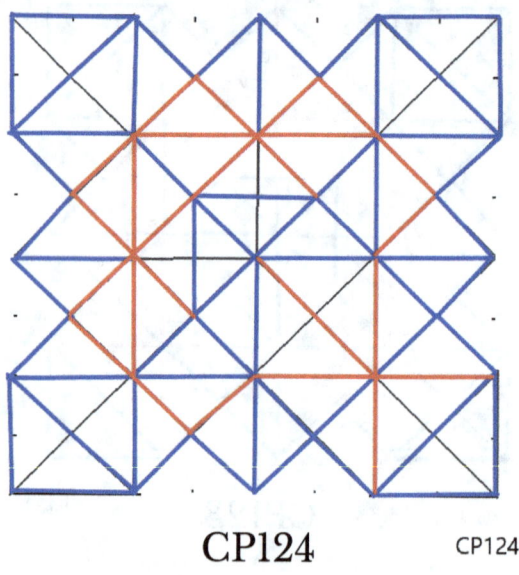

CP124   CP124

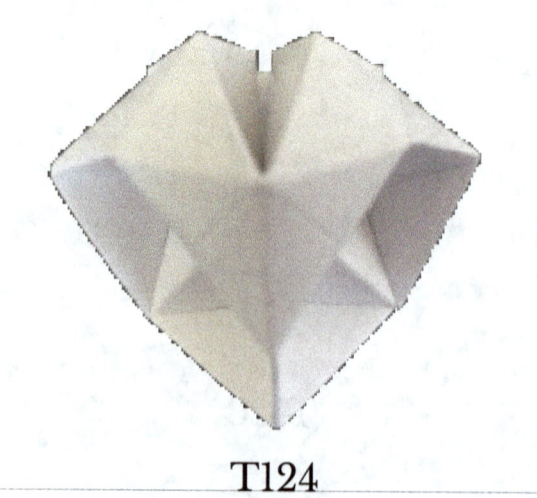

T124

T125

Model T125 has two Niches eyes in front side and a x-hole and two squares set in the back side.

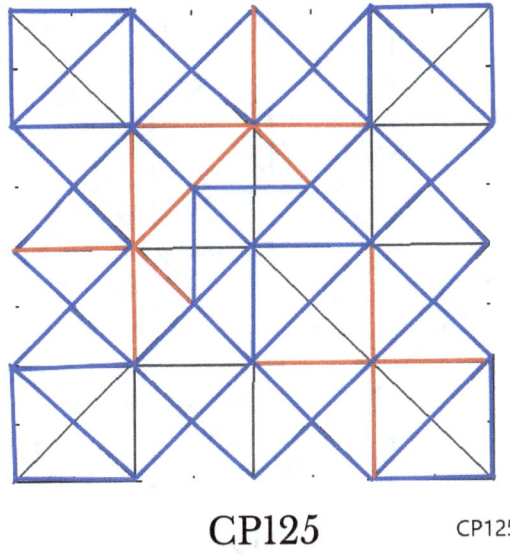

CP125   CP125

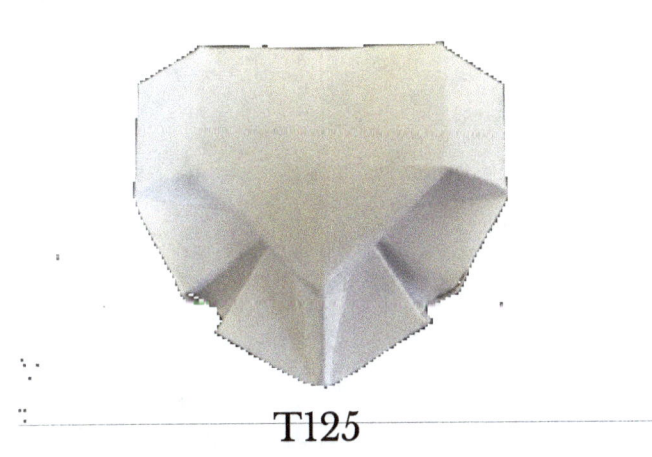

T125

T126

Model T126 has two Niches eyes in front side and a x-hole and two squares set in the back side.

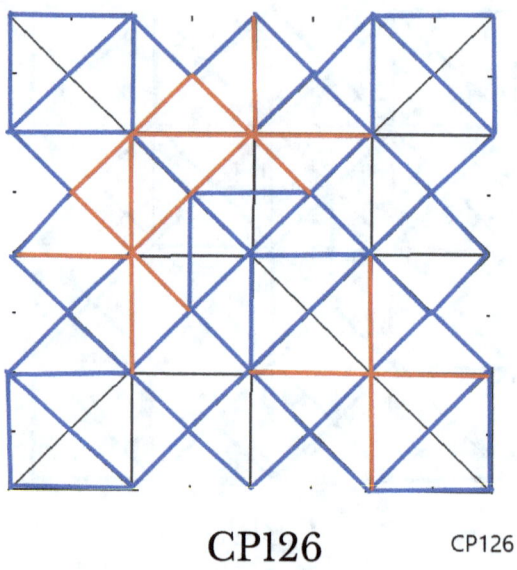

CP126   CP126

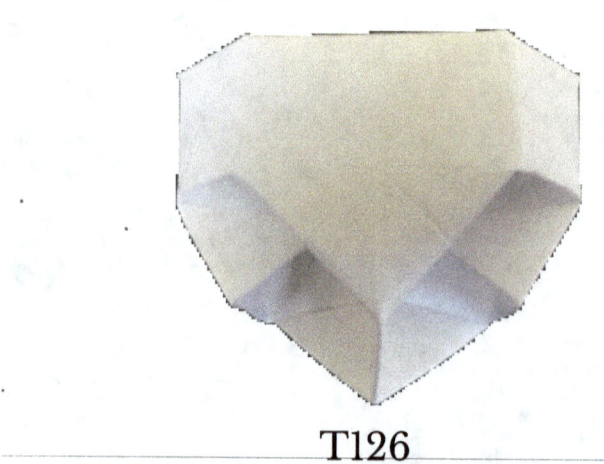

T126

T127

Model T127 has two Niches eyes in front side and a x-hole and two squares set in the back side.

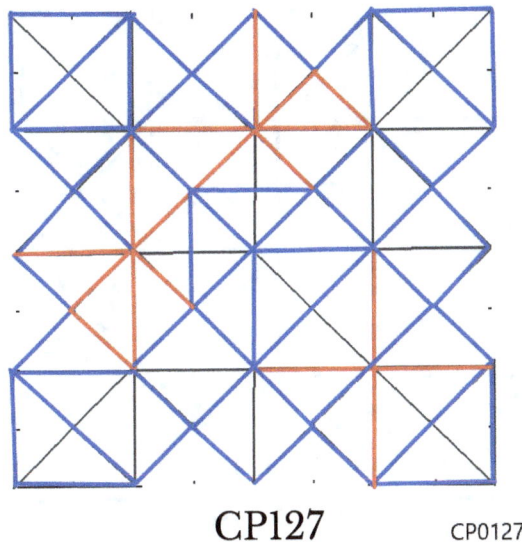

CP127   CP0127

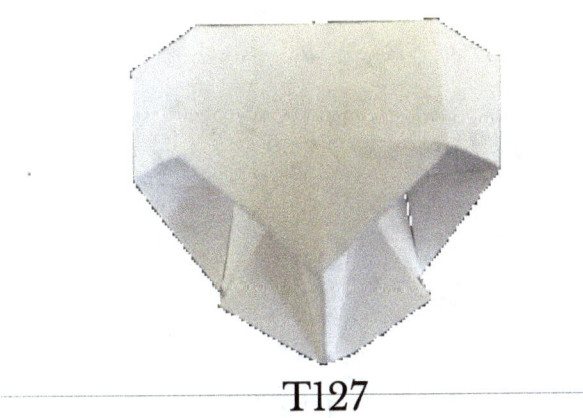

T127

T128

Model T128 has two Niches eyes in front side and a x-hole and two squares set in the back side.

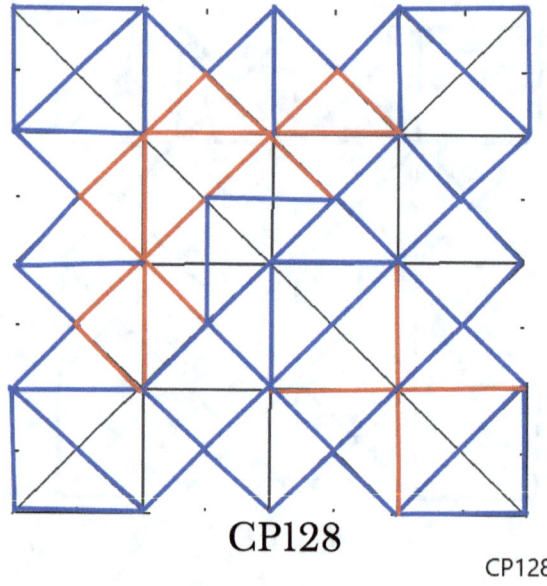

CP128

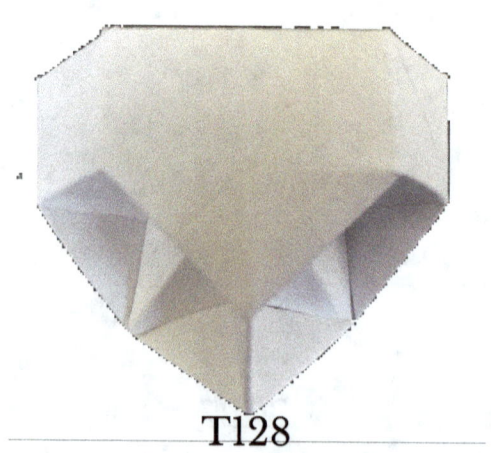

T128

T129

Model T129 has two Niches eyes in front side and a four triangles set in the back side.

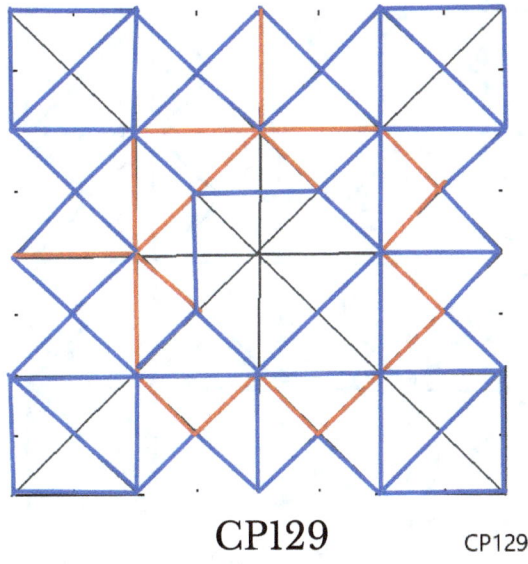

CP129   CP129

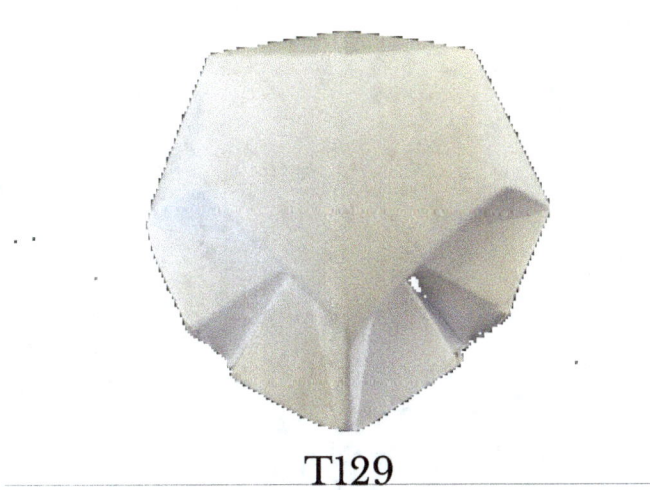

T129

T130

Model T130 has two Niches eyes in front side and a four triangles set in the back side.

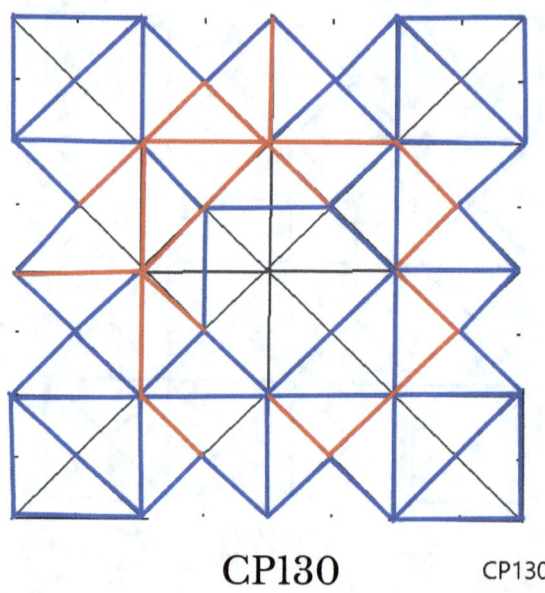

CP130   CP130

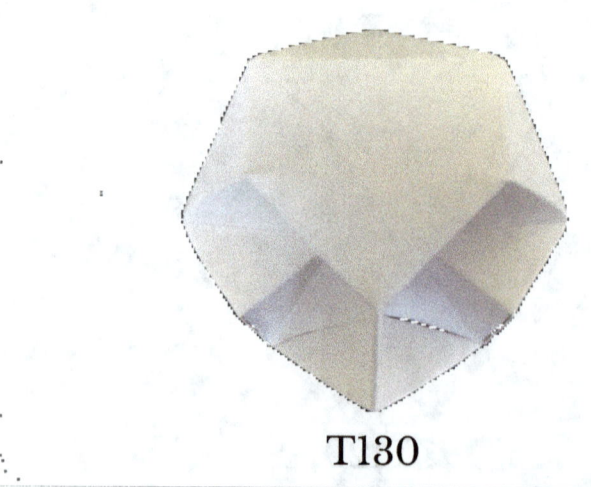

T130

T131

Model T131 has two Niches eyes in front side and a four triangles set in the back side.

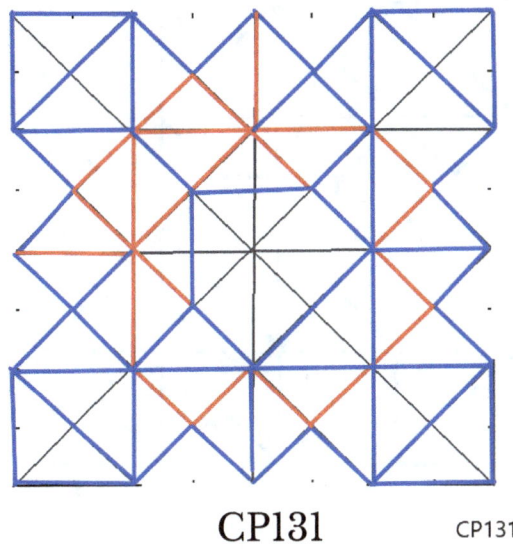

CP131   CP131

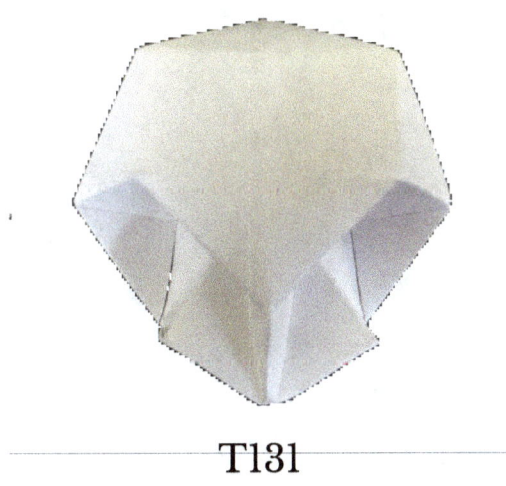

T131

T132

Model T132 has two Niches eyes in front side and a four triangles set in the back side.

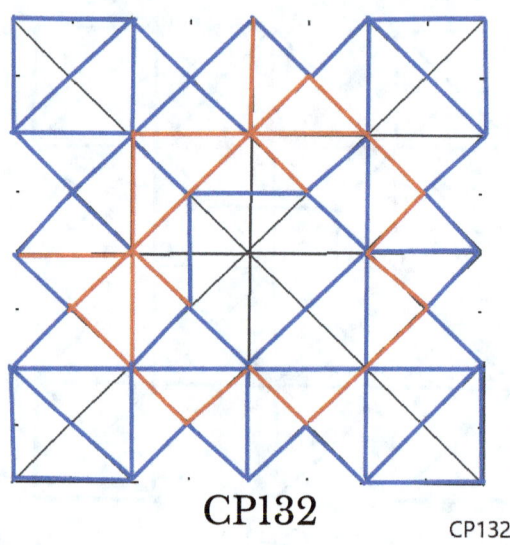

CP132

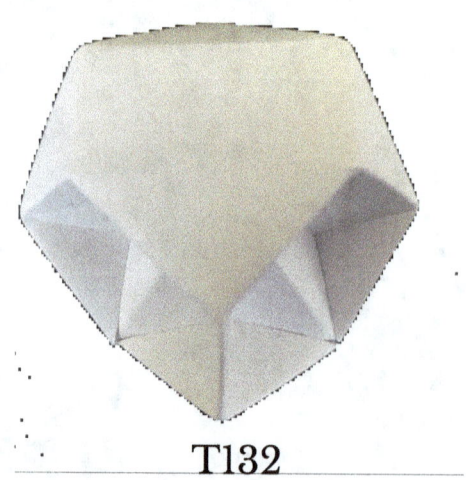

T132

T133

Model T133 has two Niches eyes in front side and a two triangles set in the back side.

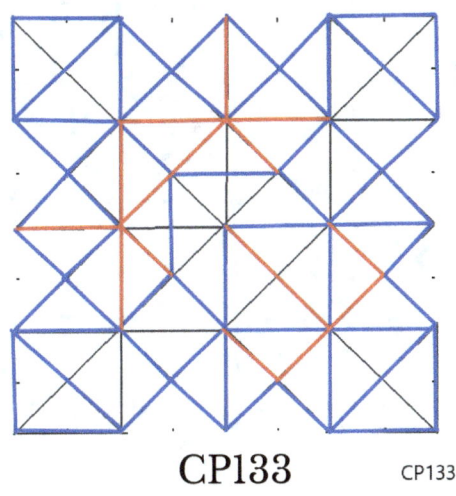

CP133   CP133

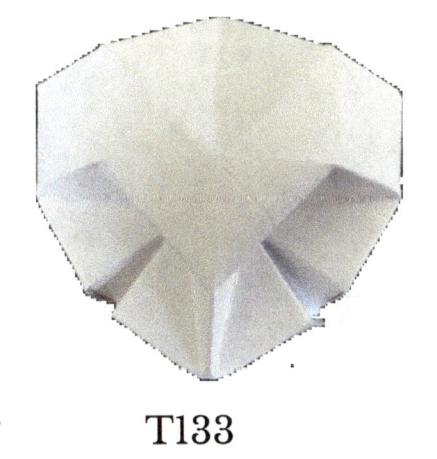

T133

T134

Model T134 has two Niches eyes in front side and a two triangles set in the back side.

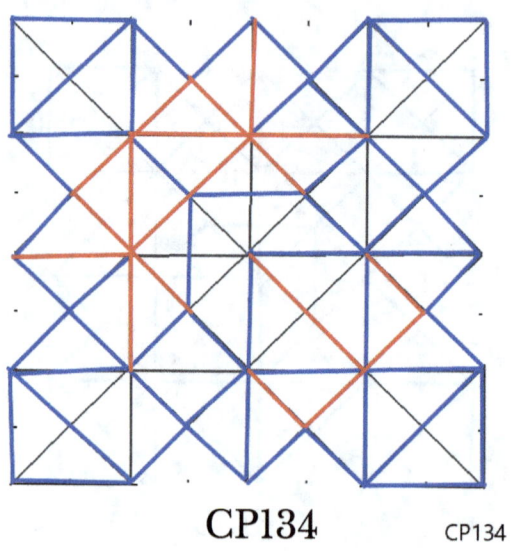

CP134   CP134

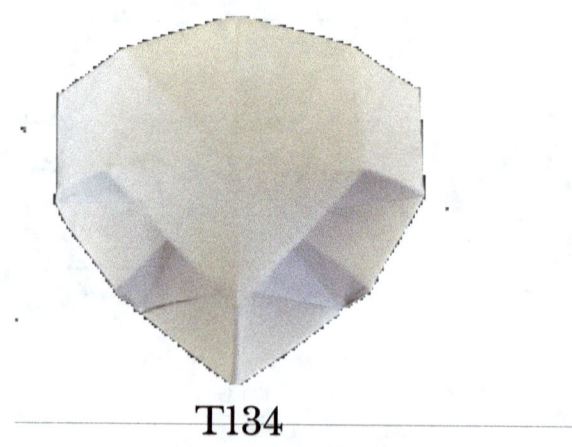

T134

T135

Model T135 has two Niches eyes in front side and a two triangles set in the back side.

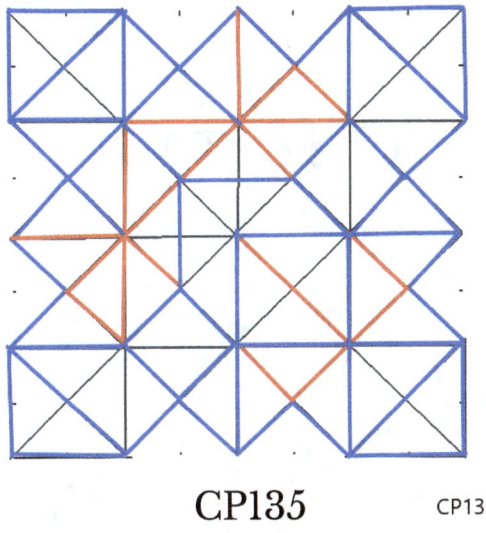

CP135   CP135

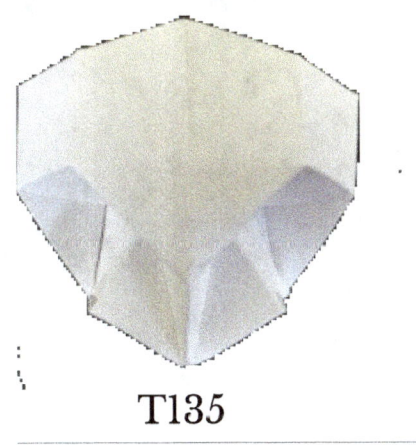

T135

T136-T150

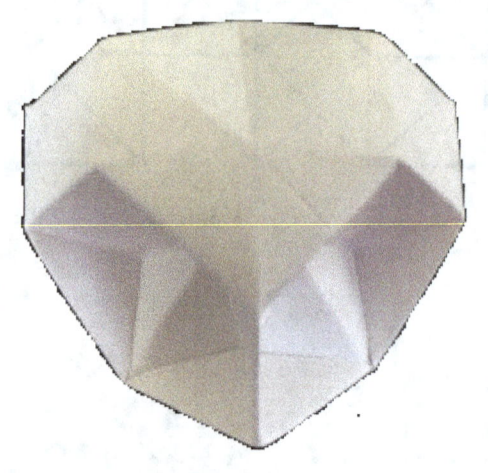

T136

Model T136 has two Niches eyes in front side and a two triangles set in the back side.

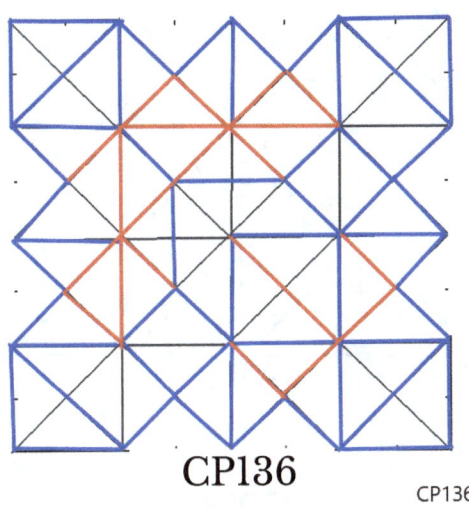

CP136

CP136

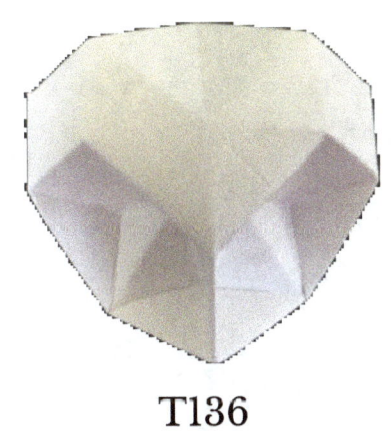

T136

T137

Model T137 has two Niches eyes in front side and a square cap in the back side.

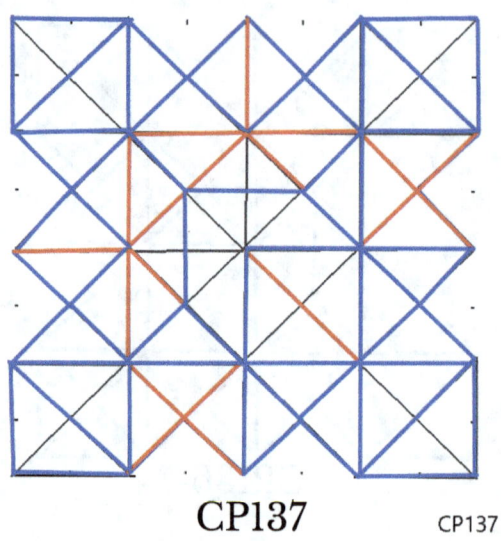

CP137   CP137

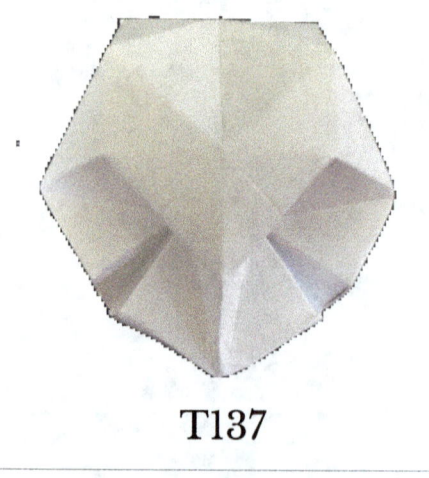

T137

T138

Model T138 has two Niches eyes in front side and a square cap in the back side.

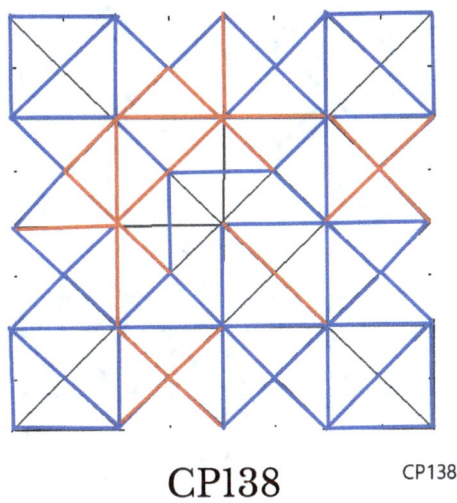

CP138   CP138

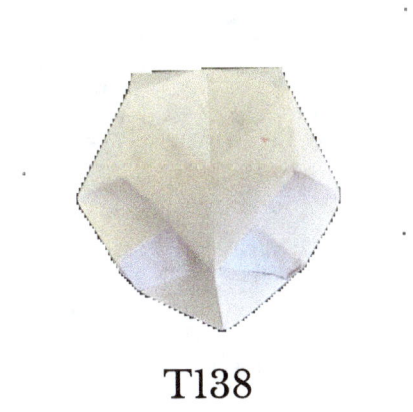

T138

T139

Model T139 has two Niches eyes in front side and a square cap in the back side.

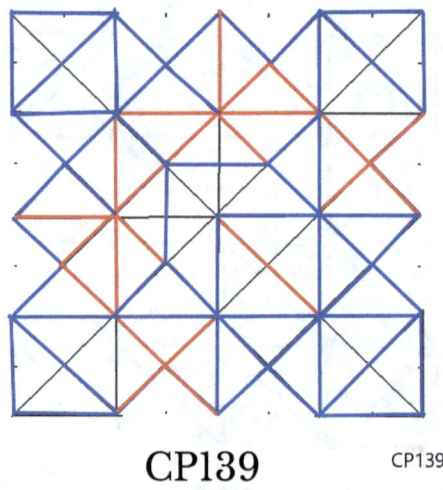

CP139 CP139

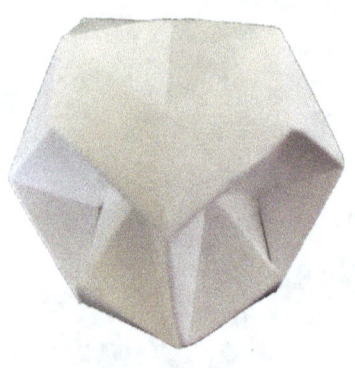

T139

T140

Model T140 has two Niches eyes in front side and a square cap in the back side.

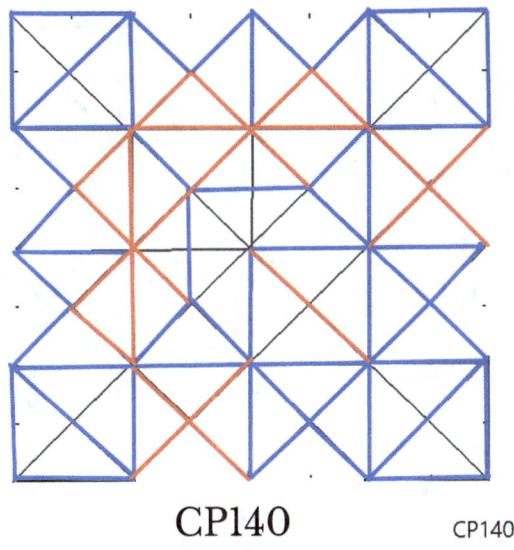

CP140 CP140

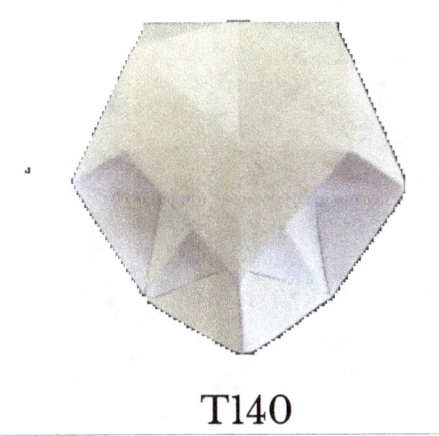

T140

T141

Model T141 has two large Niches eyes and two ears. A pair of Hinge creases applied.

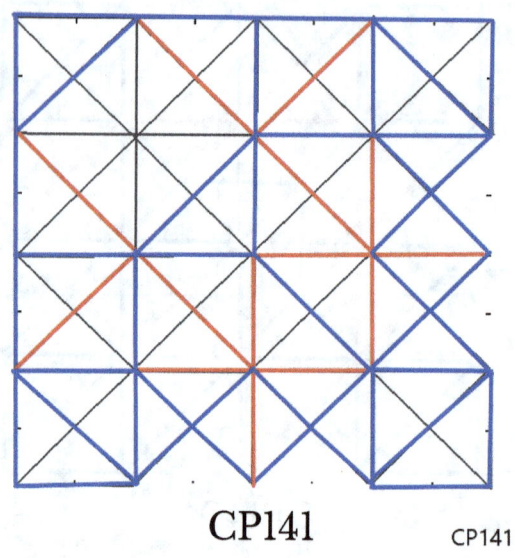

CP141

CP141

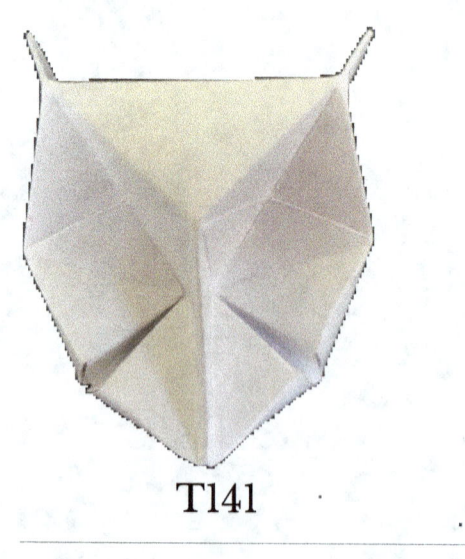

T141

T142

Model T142 has two large Niches eyes and two ears. A pair of Hinge creases applied.

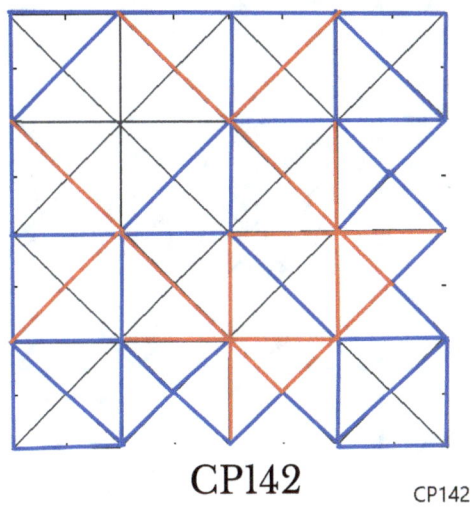

CP142

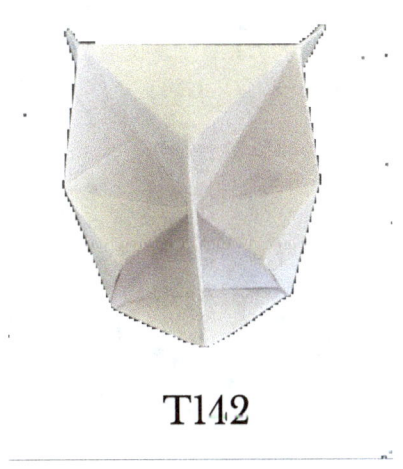

T142

T143

Model T143 has two large Niches eyes and two ears. A pair of Hinge creases applied.

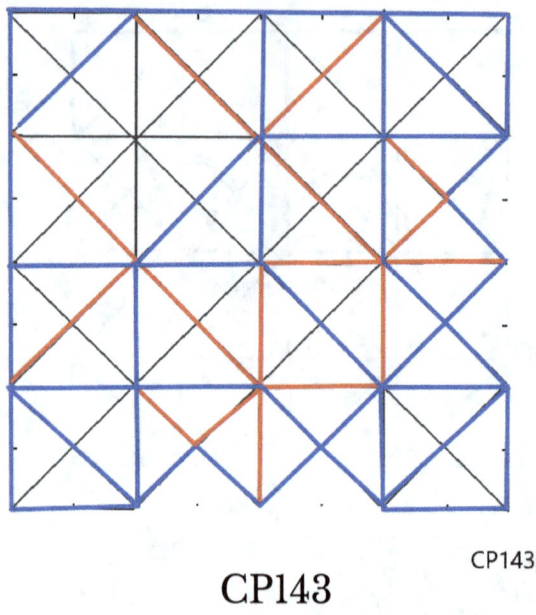

CP143

T143

T144

Model T144 has two large Niches eyes and two ears. A pair of Hinge creases applied.

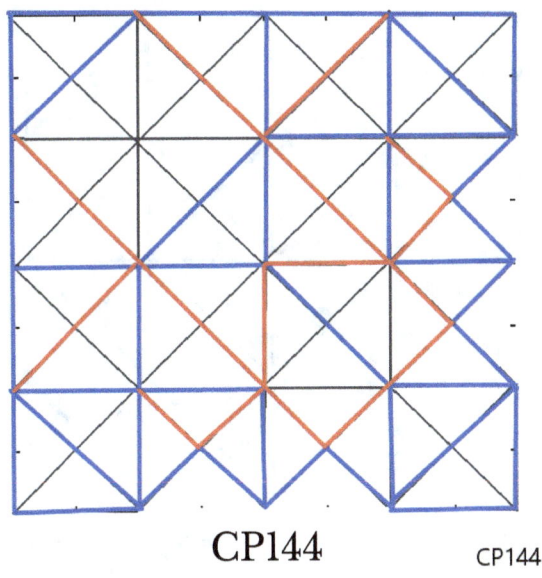

CP144   CP144

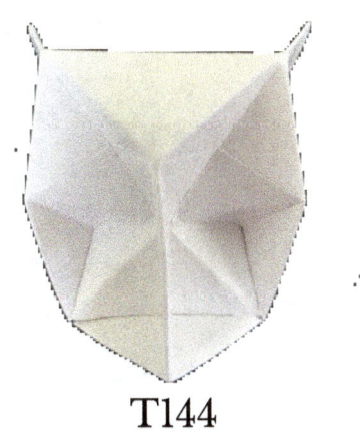

T144

T145

Model T145 has a large nose and mouth with moustache. A pair of Hinge creases applied.

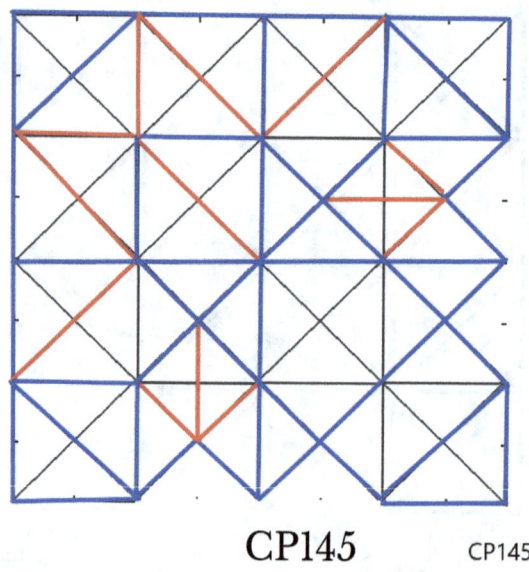

CP145   CP145

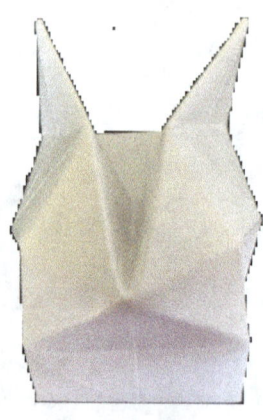

T145

T146

Model T146 has a two-diamond set moustache and eyes. A pair of Hinge creases applied.

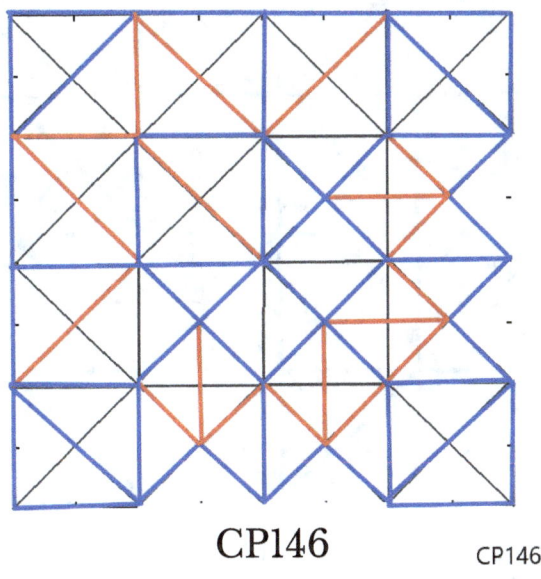

CP146   CP146

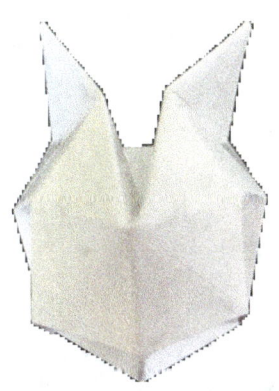

T146

T147

Model T147 has two sad eyes and two rabit ears. A pair of Hinge crease applied.

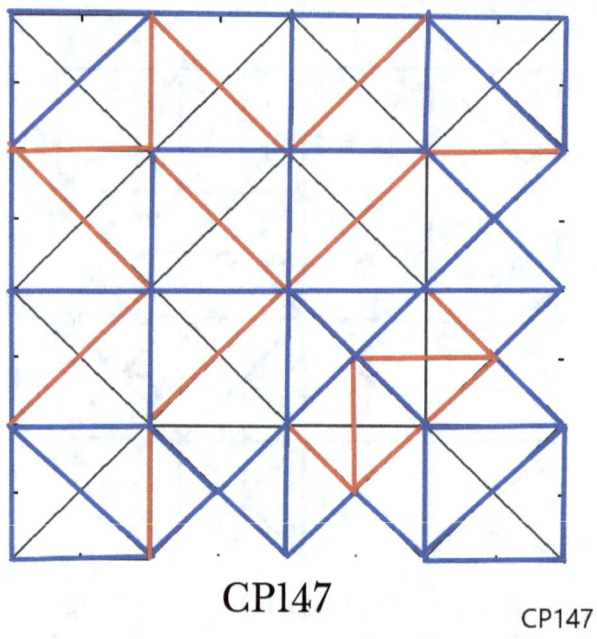

CP147

CP147

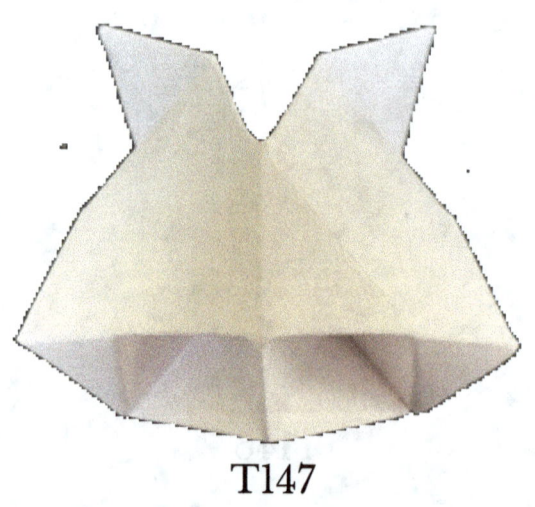

T147

T148

Model T148 has two sad eyes and two rabit ears. A pair of Hinge crease applied.

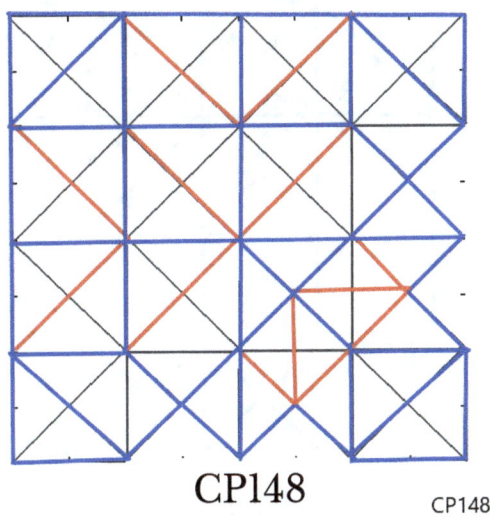

CP148

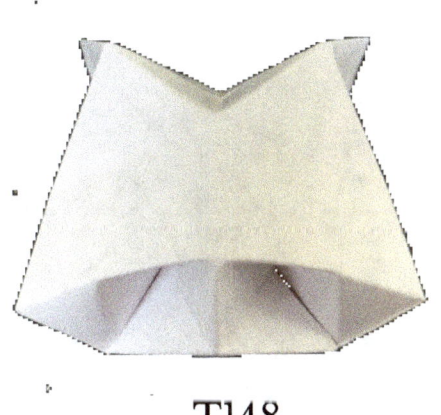

T148

T149

Model T149 has a diamond-triangle nose and mouth set and two rabit ears. A pair of Hinge crease applied.

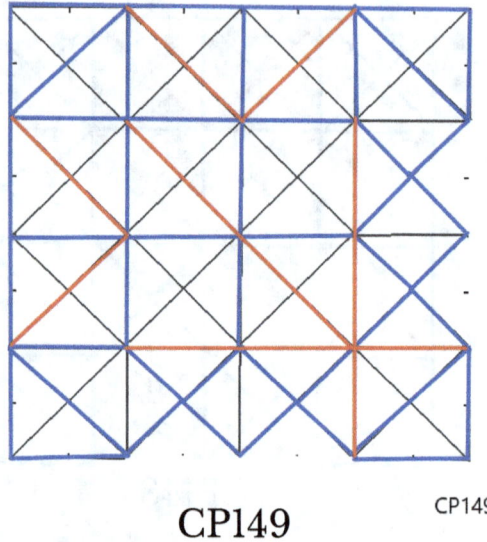

CP149   CP149

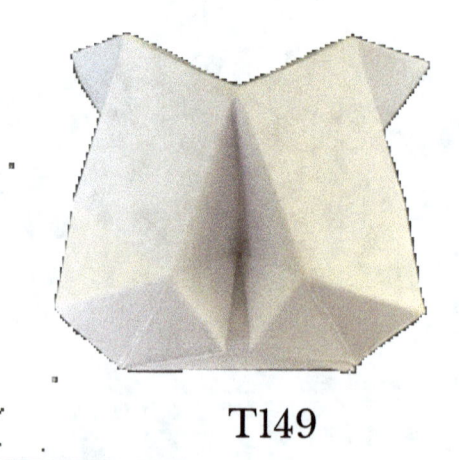

T149

T150

Model T150 has a diamond-triangle nose and mouth set, two triangle eyes, and two rabit ears. A pair of Hinge crease applied.

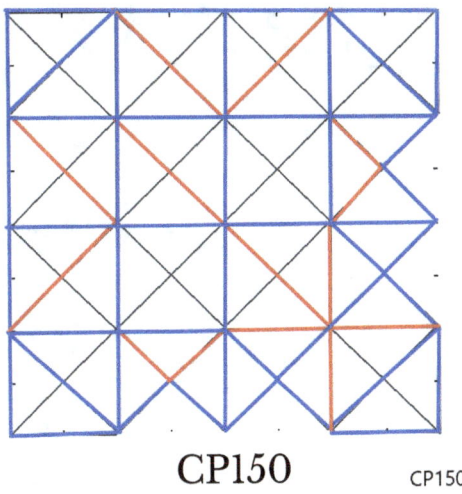

CP150

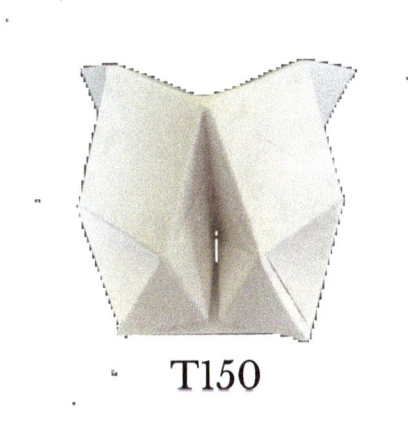

T150

www.ingramcontent.com/pod-product-compliance
Lightning Source LLC
Chambersburg PA
CBHW072053110526
44590CB00018B/3155